Praise for

Essential Rammed Earth Construction

Rammed earth construction is enjoying a renewal as people recognize it as a great, climate-friendly way to build. With research and experimentation all over the world expanding the range of clay construction, Tim Krahn brings a much-needed and science-based update to a North American audience of designers, engineers and builders. Tim, what took you so long?

— Bruce King. P.E., author, *The New Carbon Architecture*

Essential Rammed Earth Construction is a great book for anyone who wants to deepen their technical knowledge of rammed earth walls systems. I appreciate all the work Tim has put in to aggregating a vast body of knowledge into a useful book. It's very helpful to have a book on rammed earth that is more focused on engineered rammed earth walls for cold climates.

— Clifton Schooley, Clifton Schooley & Associates, Rammed Earth Designers and Builders

This book provides excellent background, engineering science and practical advice for constructing rammed earth structures, from an established Canadian designer and builder. Tim Krahn takes us through the whole story, touching on current research into the mechanical and thermal properties of the material, to testing for compliance, to guidance on methods and details. The guide provides a welcome, up-to-date view of rammed earth construction and will be especially of interest to those building in colder climes.

— Charles Augarde, Professor in Civil Engineering, Durham University, UK

Rammed Earth as a building technique has been practiced for around 10,000 years. It is arguably the most popular method of building worldwide. The demands of modern building require modern information. This book provides it. Both old and new information and techniques are covered so very well. With excellent advice for professionals and for beginners alike, it is an excellent book that should expand the successful use of rammed earth worldwide, thereby making the world a better place.

— Stephen Dobson, Ramtec, Western Australia. Builder of over 750 rammed earth structures since 1976.

Tim has produced an excellent, well-balanced book. *Essential Rammed Earth Construction* will provide a comprehensive guide for engineers, builders, architects and clients to the specifics of rammed earth construction. Thoroughly researched, both academic and practical, this book raises the bar for rammed earth construction and will prove a valuable addition to the arsenal of rammed earth designers and builders around the world.

— Dr Paul Jaquin, Structural Engineer, Queenstown, New Zealand

One of the traditional criticism of earthen structures is that they cannot hold in harsh climatic conditions. With this excellent work, Tim Krahn presents in a comprehensive, scientifically sound and not sugar-coated way how rammed earth buildings can hold in cold climates like Canada too. The book is very useful not only for practitioners, but also for scientists and whoever is wondering if rammed earth buildings could be an effective solution to tackle climate change.

— Alessandro Arrigoni, Postdoctoral Researcher,
Dept. of Civil Engineering, University of Toronto

Tim's *Essential Rammed Earth Construction* pulls apart the subject and gets into all the interesting aspects of the material. His thoughts on rammed earth's larger-than-expected carbon footprint and potential alternatives to Portland cement give me hope for this great material as we move to lower carbon options. If you are an architect, engineer, or contractor, this book has the details to help you understand rammed earth.

— Terrell Wong OAA, President of Passive Buildings Canada,
Stone's Throw Design Inc., Architect for the Environment

Tim presents the engineering understanding of rammed earth construction in the honest voice of an experienced expert. Be you technical or practical, experienced or novice, this book has something for everyone.

— Dr Christopher Beckett, The University of Edinburgh

Essential Rammed Earth Construction has all of the essential knowledge for completing a successful rammed earth project. Written by a geo-technical engineer with experience ramming earth, the book will assure you that all of the nitty gritty details are covered.

— Kelly Hart, author, *Essential Earthbag Construction*

sustainable building essentials

essential RAMMED EARTH CONSTRUCTION
the complete **step-by-step** guide

Tim Krahn, P. Eng.

New Society Sustainable Building Essentials Series

Series editors
Chris Magwood and Jen Feigin

Title list

Essential Hempcrete Construction, Chris Magwood

Essential Prefab Straw Bale Construction, Chris Magwood

Essential Building Science, Jacob Deva Racusin

Essential Light Straw Clay Construction, Lydia Doleman

Essential Sustainable Home Design, Chris Magwood

Essential Cordwood Building, Rob Roy

Essential Earthbag Construction, Kelly Hart

Essential Natural Plasters, Michael Henry & Tina Therrien

Essential Composting Toilets, Gord Baird & Ann Baird

Essential Rainwater Harvesting, Rob Avis & Michelle Avis

Essential Rammed Earth Construction, Tim Krahn

See www.newsociety.com/SBES for a complete list of new and forthcoming series titles.

THE SUSTAINABLE BUILDING ESSENTIALS SERIES covers the full range of natural and green building techniques with a focus on sustainable materials and methods and code compliance. Firmly rooted in sound building science and drawing on decades of experience, these large-format, highly illustrated manuals deliver comprehensive, practical guidance from leading experts using a well-organized step-by-step approach. Whether your interest is foundations, walls, insulation, mechanical systems, or final finishes, these unique books present the essential information on each topic including:

- Material specifications, testing, and building code references
- Plan drawings for all common applications
- Tool lists and complete installation instructions
- Finishing, maintenance, and renovation techniques
- Budgeting and labor estimates
- Additional resources

Written by the world's leading sustainable builders, designers, and engineers, these succinct, user-friendly handbooks are indispensable tools for any project where accurate and reliable information is key to success. GET THE ESSENTIALS!

Copyright © 2019 by Tim Krahn.
All rights reserved.

Cover design by Diane McIntosh.
Cover images: Bottom right: Emily Blackman. Top middle and top right: Tim Krahn.
Illustrations by Dale Brownson.
All uncredited photos in the book: Tim Krahn.
Chapter header image: © Emily Blackman. Sidebar background: Tim Krahn.

Printed in Canada. Second printing March 2022.

This book is intended to be educational and informative. It is not intended to serve as a guide. The author and publisher disclaim all responsibility for any liability, loss or risk that may be associated with the application of any of the contents of this book.

Inquiries regarding requests to reprint all or part of *Essential Rammed Earth Construction* should be addressed to New Society Publishers at the address below. To order directly from the publishers,
please call toll-free (North America) 1-800-567-6772, or order online at www.newsociety.com

Any other inquiries can be directed by mail to:
New Society Publishers
P.O. Box 189, Gabriola Island, BC V0R 1X0, Canada
(250) 247-9737

LIBRARY AND ARCHIVES CANADA CATALOGUING IN PUBLICATION

Krahn, Tim, 1971-, author
    Essential rammed earth construction : the complete step-by-step guide / Tim Krahn.

(Sustainable building essentials)
Includes bibliographical references and index.
Issued in print and electronic formats.
ISBN 978-0-86571-857-9 (softcover).--ISBN 978-1-55092-651-4 (PDF).--
ISBN 978-1-77142-246-8 (EPUB)

    1. Pisé--Handbooks, manuals, etc. 2. Earth construction--Handbooks, manuals, etc. I. Title. II. Title: Rammed earth construction. III. Series: Sustainable building essentials

TH1421.K73 2018   693'.22   C2018-905503-0
                                                                                         C2018-905504-9

New Society Publishers' mission is to publish books that contribute in fundamental ways to building an ecologically sustainable and just society, and to do so with the least possible impact on the environment, in a manner that models this vision.

Contents

Acknowledgments		ix
Foreword	By Meror Krayenhoff	xi
Chapter 1:	Introduction	1
Chapter 2:	Rationale and Appropriate Use	3
Chapter 3:	Building Science Notes	9
Chapter 4:	Materials	21
Chapter 5:	Wall System Examples and Structural Design Considerations	53
Chapter 6:	Tools and Mixing	71
Chapter 7:	Construction Methods	81
Chapter 8:	Cost Estimates Based on a Cement-Stabilized Rammed Earth Project	101
Chapter 9:	Wall Surfaces, Openings, and Embedments	105
Chapter 10:	Building Code Developments	119
Appendix A:	Sample Engineering Specification	125
Appendix B:	Alternative Solutions Proposal	131
Definitions		133
resources and material suppliers		135
Bibliography		137
Index		139
About the Author		145
A Note About the Publisher		146

Acknowledgments

I WOULD LIKE TO THANK CHRIS MAGWOOD AND JEN FEIGIN for asking me to be one of the Sustainable Building Essentials authors. I enjoy working with them as a teacher, designer, and builder — and also as a friend.

I would also like to thank my mentor, friend, and business partner, Kris Dick, who has helped me continue to develop as a professional and as a person. And Mark West, who taught me about the value of improvisation and finding invention through rigor.

I am very grateful for the support and patience of my wife, Dalila Seckar, who puts up with a work-from-home engineer who has trouble saying "no."

I cannot possibly put together a comprehensive list of key figures in earthen construction across the globe, but I would be remiss not to mention those who have directly and indirectly influenced me and my work. Gernot Minke in Germany, Pete Walker in the UK, builders and researchers at CRAterre in France, Venkatarama Reddy at the Indian Institute of Science, Martijn Schildkamp at the Auroville community in India, and many other contemporary builders are successfully creating durable raw earth structures. Martin Rauch, an Austrian architect, has made tremendous headway in using textures to minimize rain-driven erosion, and he has promoted the use of a sacrificial exterior layer to expand the vocabulary of raw rammed earth building.

Modern earthen construction in Australia, New Zealand, China, Canada, and the US is predominantly of the stabilized variety. Builders like Steve Dobson in Australia, David Easton in the US, and Meror Krayenhoff in Canada have made considerable progress in using stabilized rammed earth as a durable, beautiful building material in both residential and commercial projects. David Easton is working on moving beyond cement stabilization with his Watershed Blocks, and this is a very promising technology for building sustainably within the current North American construction context.

In recent years, modern earthen construction has been investigated from a geotechnical engineering point of view, and the work of Charles Augarde, Paul Jaquin, Matthew Hall, Pete Walker, Venkatarama Reddy, Jean-Claude Morel, Christopher Beckett, and Daniela Ciancio — among others — is helping bring current applied science and engineering analysis techniques to bear on this ancient material. I encourage readers who are interested in the potential of both raw and stabilized rammed earth to look at the work of these people and their organizations. It is inspiring on many levels.

I want to thank James Blackman for his help conveying the practical matters of building with rammed earth. And last but not least, thanks to Rob West and Linda Glass for their editorial expertise.

Foreword

Written by Meror Krayenhoff

THE IDEA OF BUILDING WITH RAMMED EARTH came to Canada in 1992, when the Sustainable Salt Spring Island group held a Sustainable Building Conference that featured speakers from around North America. We were inspired by notions like healthy buildings, permaculture, underground buildings, straw bale buildings, and the highlight — rammed earth buildings. Many were inspired by David Easton, a prominent figure in the rammed earth world, but was it possible in the Canadian climatic context? Twenty-six years later we have the answer: there are rammed earth buildings in most Canadian provinces. Was it easy? No, almost everything had to be reinvented.

The beginning of Canadian rammed earth took place in the most challenging location. At the edge of the rain forest with frequent horizontal driving rain, in the highest North American seismic zone at the north end of the San Andreas fault, and with freeze/thaw cycles that can number up to 30 in one 24-hour period, this was not a location where rammed earth solutions used elsewhere on the planet could be duplicated.

That reinvention in 1992 began with weekly brainstorming meetings that included an engineer, an architect, a senior building inspector, a formwork designer, and myself, an experienced home builder. Our first success was a rammed earth addition on our home in 1993 — the first Code-approved rammed earth project in Canada.

For the first 12 years we were the only rammed earth builder in Canada, and the technology was not well understood. We were viewed as a marginal voice in the wilderness. It became clear that this rammed earth technology could not have the environmental impact that the planet so needed if we carried on as we were.

So we began teaching weekend courses, then week-long courses, and then we tried franchising (which didn't work out so well). The net result was that now there were rammed earth builders other than us spreading the word and further developing the technology.

Ten years ago I used to say that everyone who was doing rammed earth in Canada got their start with us. Now there are Canadian rammed earth builders that I've never heard of making their own improvements to the technology, which is inspiring. And this technology is also being used and developed in at least another 20 countries. A momentum is building, and this book will go a long way to accelerating that momentum.

It's a different kind of momentum. In the practical and regulatory side of building, the way things develop is almost exclusively reactive. The evolution of the Building Code in my lifetime has been a process of looking in the rear view mirror and trying to fix what's not working. This whack-a-mole approach has resulted in wood-frame buildings that have toxic interior environments and life expectancies of 40 years or so. We have not *reacted* ourselves into a healthy and durable housing stock. We can and must do better as this historical approach cannot anticipate the climate-changed world that buildings built today will need to endure.

In almost everything we do, we have the choice to react or create. What if we were to *create* a really great way to build, completely ignoring the inertia of today's building norms? I like to imagine a global building stock that doesn't require energy to heat or cool, with zero toxic chemicals outgassing. The maintenance is almost zero. Every home is unique and beautiful. There is a visual and visceral connection to nature. The humidity flywheel effect (where the rammed earth absorbs excess water vapor and releases it later) ensures there is no mold. Even if climate change brings 24" of rain down in a few hours (as we are now seeing in a variety of locations), the home is barely affected. Where fires used to burn down suburbs, the houses now stand unaffected (if they have green roofs and other fireproof details). Where hurricanes and twisters demolished large suburban areas, the houses now remain intact, acting as above-ground hurricane shelters. When there are heat waves, people can take refuge in their homes without relying on air conditioning. The materials for this housing stock can be reused for the same purpose (*Cradle to Cradle*), thereby almost eliminating resource extraction and landfilling. The multifamily portion of this future global housing stock ensures acoustical sanctuary for neighbors who are side by side, as well as from outside noise. Houses last for 2,000 years, like the Pantheon. Surely we can build as well as the ancient Romans.

This kind of global housing stock would do wonders for the global economy and our resilience. According to the US Green Building Council, buildings account for 40% of US energy consumption. Canada's consumption is similar, and that is really low-hanging fruit when it comes to saving energy. The cheapest watt is the watt that is not needed — far cheaper than supplying energy with solar and wind, as those options, again, involve disposability.

But won't that cost a lot of money? And what about the embodied energy and carbon footprint of the cement? These are the questions that all rammed earth builders face repeatedly. Here's how I answer. If I were to suggest that we should all be eating with plastic spoons because they're way cheaper than metal spoons and have far less embodied energy, almost everyone would understand that although the metal spoon costs 100 times more than the plastic spoon, it is worth it for its durability, its beauty, and environmental appropriateness — in short, its life-cycle benefits. Quality is often like that.

We have become so accustomed to and accepting of our disposable culture that we barely see the benefits of durability. It's the disposability of the things we consume that keeps us going back to Mother Nature for more materials and keeps us expanding our landfills and rendering our aquifers toxic. Disposability is arguably the single biggest blight on the environmental health of our planet. When the dominant question is "How much does it cost?," we are effectively embracing disposable products while ignoring the cost to the planet and our descendants.

We need to be rigorous in talking about life-cycle costs. Similarly, the initial embodied energy conversation is myopic and harmful to the planet. How much embodied energy is in the many stick-framed houses that need to be rebuilt in the lifespan of one well-built rammed earth house? That is a life-cycle embodied-energy question, and the outcome of that analysis is so different (rammed earth life-cycle embodied energy is stellar). We desperately need to be talking about life-cycle everything, not just because rammed earth looks good in this light, but because the viewpoint of the future as someone else's problem is why we are in this environmental mess.

The two frequent criticisms of rammed earth (initial cost and initial embodied energy) open the door to conversations that can incorporate longer-term thinking. Long-term thinking is an expression of care for our shared future. Actions based on life-cycle thinking honor our children and grandchildren out to and including seven generations. The beauty of rammed earth expresses that kind of love.

I believe that the *created* future, where we get everything we want in buildings is not only possible, but not far off. Much of the wish list for building is already possible through the use of quality rammed earth, and the way to implement it is clearly laid out in this book, which can be thought of as a guide book to a better future.

Quality rammed earth housing has the capability to transform the experience of being human. Currently, we think that houses are expensive to build and to maintain. The temperature and humidity can swing outside our comfort zone. We frequently cohabit with rodents, molds, carpenter ants, termites, powderpost beetles, and many other insects that make their home inside easily accessible, warm and dry wood-framed walls. Our buildings are vulnerable to fire, flooding, and very high winds. Being touched by nature in those ways is unpleasant, and actually unnecessary.

How to avoid most of these hazards is fairly easy to understand when one begins to see what the rammed earth house can offer. The initial building cost is perhaps the hardest to understand and how to implement a reduction in cost to the initial home buyer will need some brainstorming. Consider a rammed earth house that lasts 200 years, and a stick-frame equivalent that lasts 40 years.

While there are ongoing attempts to increase the durability of stick-framed homes, the difficulty that won't go away is that organic materials are designed by nature to decompose. Wood can be toxified with arsenic, lead paint, fungicides, and a host of other biocides, but the uncomfortable truth is these toxify our environment and typically don't last more than a decade or two. The key variable in this equation is if the rammed earth can last for 200 years, and whether it can withstand the wet/dry cycling and freeze/thaw cycling over that time frame. That is a function of its density and compressive strength.

The above example of a 200-year life cycle, which shows the rammed earth home at ¼ the life-cycle cost of the stick-frame home, does not factor in the reduced heating/cooling costs, reduced maintenance, landfilling costs, and health care impacts. When those are factored in, the life cycle cost of a rammed earth home could be 15%–20% of the stick-frame home. If we can only figure out how to bring at least some of those savings to the initial rammed earth home buyer, then we will have unlocked the door to this industry.

One of the things that we have found is that while most homeowners are not willing to look at anything with longer than a 3–5 year payback, governments are willing to take the long view. They know that they will be paying for the heating and cooling, the maintenance, the employee sick days, the pest control, the rebuild, and the landfilling. They find life-cycle costs very interesting, and are often willing to cut square footage to achieve operating savings. They also understand the political message that is delivered by using local materials and local labor.

Many inspired and dedicated people have contributed to the development of insulated rammed earth and the state of the residential technology is well captured in this book. Gathering as much material as Tim has is an enormous task, and necessarily a labor of love. Everyone who builds with insulated rammed earth owes Tim a debt of gratitude.

Building with rammed earth is easy, but building it well is a challenge. This book is a must-read for anyone considering building with rammed earth, and given that Canadian insulated rammed earth technology is by far the best in the world, it is mandatory reading for all insulated residential rammed earth building done anywhere on the planet.

	Stick frame	Rammed earth
Initial Cost	IC	1.25 × IC
Life Cycle	40 years	200 years
Life Cycle Cost	5 × IC	1.25 × IC

"Tamp 'em up solid, so they won't come down."

— unknown lyricist

Chapter 1

Introduction

THERE ARE MANY appealing features to building with rammed earth. Aesthetically, rammed earth is very pleasing — from the sedimentary layer effect of the lift lines to the surface textures on a wall. Site materials can be used to create major structural and building envelope elements, which means low embodied energy and a small carbon footprint for those components. Rammed earth is inherently massive, which translates into interior thermal stability, even when there are large temperature swings outdoors (especially within an insulated envelope). The mix design generally includes clay, so it has an open pore structure that, depending on the application of sealers, allows rammed earth to absorb and shed water vapor, which can modulate extremes in indoor humidity. Raw rammed earth construction can be carried out using entirely nontoxic materials, fostering a healthy indoor environment with no added volatiles or toxins in the air. Stabilized rammed earth can act as both an interior and an exterior finished wall surface, even in harsh northern climates. This means minimal long-term maintenance because it eliminates the need for paints and stains — although unstabilized mixes on exterior walls may require periodic application of sealers, a protective plaster coating, or even a rainscreen assembly for extreme conditions.

I am currently a practicing professional engineer. While I first came to construction over 30 years ago, it was as a laborer, then a carpenter, and then an amateur mason. Now I collaborate closely with many experienced builders and aspiring ones (along with owner-builders, who usually fall somewhere between the two), but I do not regularly hold tools in my hands other than a computer mouse, a calculator, and pen and paper. I do continue to build things myself as a hobby and for research purposes, and where possible I do like to get my hands dirty on job sites — but the majority of my time is spent at my desk, not on site. That said, this book is directed primarily to builders.

Many professionals in the rammed earth building community have contributed to this book, allowing me to present readers with the current state of the art.

This book is unique because its approach is from a North American point of view, in particular Canada and the northern US, where the cold climate requires additional insulation to be incorporated into the building envelope. Freeze-thaw cycles require considerably more attention in both materials and detailing. High snow loads are common, and wind and seismic loads are also prevalent. Canada also has relatively conservative, limit-states design codes for structural engineering. This book will review several international codes and discuss the ramifications for builders working in Canada and the US.

I am often reminded of a conversation I had with George Nez, a pioneer of thin-shell roof construction. He had traveled up to southern Ontario from his home in Colorado to help run a workshop on how to build roof elements with various fabrics and acrylic-cement-sand mixes applied in layers on first-order hyperbolic shapes. He had been observing how the builders, students, and designers responded to his techniques for several days. We were together in the shade on a hot day watching students apply a second layer to one of the forms. While everyone involved agreed that this was a novel method, most of us were trying to imagine ways we could make these roofs work with our own building modes. In an almost exasperated voice, he told me that we were all "wall builders." He basically thought we were missing the point. The people in the area in Africa that George had been working in during the 1960s needed *overhead* shelter far more than they needed walls. The principal reason for developing this method had been the need for lightweight, durable roofs that could be either built in place or lifted onto simple pole structures. If walls were desired, with this method, they could be in-filled later. But for many reasons (cold weather and swarming insects being the first that come to mind), builders in my part of the world do, indeed, tend to be wall builders first and roof builders second. So, this book begins by examining where and when rammed earth is appropriate, focusing on walls (Chapter 2).

From there the topic shifts to focus on design considerations and building science. The four control layers — water, air vapor, and thermal — are each discussed in detail (Chapter 3).

Consideration of the materials involved in rammed earth follows, including examination of the properties and role of clay, stabilizers, aggregate, and sealers. Appropriate on- and off-site testing is discussed in detail (Chapter 4).

The structural criteria for raw and stabilized rammed earth buildings are covered in Chapter 5. Some of the topics included are wall height/thickness ratios, loads and stresses, wall length limits, openings and attachment points, and provision for utilities. Section and elevation drawings of several wall systems are presented.

After characterizing the material, a discussion of necessary tools and labor follows, covering both state-of-the-art industrial methods and low-tech, pre-industrial techniques (Chapter 6).

A range of formwork options are presented in Chapter 7, and details regarding insulation, corners, different construction configurations, and workflow are discussed. General tips and techniques and instruction about removing formply are given, and a word about volunteer labor is included.

Chapter 8 gives cost estimates based on a 2015 project using cement-stabilized rammed earth with interior insulation. Ranges of costs for materials, design, equipment, and labor are given.

Finishes, maintenance, and repairs are covered in Chapter 9.

Finally, a survey of existing codes, testing standards and building permit considerations is presented. A sample specification is given, as well as an example of an alternative solutions proposal from a recent Canadian project (Chapter 10).

A word about units: I will use both Imperial and SI units in this text, as practicing engineering in Canada brings with it a need to be "bilingual" in terms of measurement. I apologize to anyone for any confusion this may cause, and I trust that we will all check our sums to avoid any errors.

There is a bibliography at the end of the book, but I give notable academic paper references at the end of some of the chapters. This book is not aiming to be a comprehensive survey of the academic literature, but my practice as an engineer is informed by current research whenever possible. The interested reader is encouraged to explore the literature — there is quite a lot of research going on in earthen construction worldwide.

Chapter 2
Rationale and Appropriate Use

I HAVE ALWAYS BEEN FASCINATED by the idea of creating a structure entirely out of material found on the site. Rammed earth is not always made up of in-situ material, but it holds that potential. Earth is a material available almost everywhere above sea level, and it is likely that the first permanent buildings were earthen. Pretty much anywhere humans have managed to build maintainable roads, there exist the basic soil elements — gravel, sand, silt, and clay (and water) — necessary to create raw rammed earth. It is because of this availability that earthen construction is among the oldest types on the planet, and it is still common in most of the world. Whether it's adobe, wattle and daub, cob, compressed earth block, or rammed earth — be it raw or stabilized — more than a billion people (as well as other animals, termites and many other insects, and many species of bird) live in earthen structures.

Definitions: Historic and Modern Additives

What is rammed earth? The name tells us both the method and the material. It is a mix of damp soil elements (earth) manually compressed (rammed) to a high density, held together by a combination of some type of binder and the effects of surface tension (what an engineer might call *matric suction*). In the literature on the subject, you will see much discussion about the desirability of *raw* versus *stabilized* rammed earth. Why differentiate between raw rammed earth and stabilized? In simple terms, raw rammed earth is made up of only gravel, sand, silt, clay, air, and water strategically mixed and then rammed into a pre-made formwork. Stabilized rammed earth adds an additional binder (usually pozzolanic or cementitious) to the clay, which changes the surface tension effects caused by drying. Pozzolanic binders are often referred to as *Roman cement*; the name comes from the volcanic soils found near the town of Pozzuoli in Italy. Modern pozzolans are ground blast-furnace slag, ash from coal-fired power plant operation, and other by-products of high-heat processes such as calcined clay. Cementitious binders are a special set of pozzolans — they are primary products of very high-heat processes. The best-known is Portland cement, but hydrated lime can also fall into this category.

Historically, casein (dairy protein) and tar were the earliest common stabilizers added to soil mixes, along with straw or other fibrous materials to add tensile capacity (although the latter is more common with adobe and cob building). Lime and naturally occurring pozzolans have been added to rammed earth for over a millennia; Portland cement and other artificial pozzolans have been added in more recent centuries.

Raw Earth Versus Stabilized Earth

Some current practitioners, particularly in Europe, promote raw rammed earth over stabilized, citing lower carbon footprint and embodied energy along with total recyclability, among other characteristics. As a builder passionate about sustainable built environments, I agree whole-heartedly. As a professional engineer practicing in Canada, I have a difficult time designing and approving stand-alone raw earth buildings. Both the physical and regulatory climates in North America make it difficult to build code-conforming, durable raw earth structures. Freeze-thaw cycles, a feature of most building sites in Canada and the northern US, make the durability of raw earth walls dubious without the addition of a protective layer on the exterior — either a plaster or some kind of rainscreen assembly. While it is physically possible to use an earthen plaster for this protective layer, and it is theoretically possible to maintain this plaster indefinitely, the code requirements in North America do not provide an easy path to approval for such a design. Further, many clients are principally attracted to rammed earth's aesthetic qualities, and covering it up is counter to that. Additional layers also add cost and complexity to the construction process. That said, there is no practical reason stating that a builder can't construct a raw earth wall and then clad the exterior with a rainscreen to create an assembly that could be defended within the *alternative solutions provisions* of the Canadian building code. I am certain that a similar strategy could be undertaken to meet the requirements of the International Building Code (IBC) in the United States for a raw rammed earth building. In short, in order to have fully exposed rammed earth on both interior and exterior surfaces of a code-conforming structure in North America, stabilization of one sort or another is mandatory. That does not mean raw rammed earth cannot be built in a code-conforming manner, but it will need some kind of external weather protection, and it may be limited to nonseismic areas as a structural element.

In terms of material selection, we will cover both raw and stabilized rammed earth. There is a significant difference in mix design between the two. My experience is primarily with stabilized rammed earth, focusing on pozzolanic binders to increase strength along with silica-based admixtures to reduce permeability in order to promote durability under repeated wet-dry and freeze-thaw load cycles. Detractors may argue that this means we are effectively creating nothing more than damp-pack concrete, and I cannot argue. It is an approach along these lines that has allowed me to obtain a building permit for a rammed earth project in one of Canada's most restrictive jurisdictions — the city of Ottawa.

While it is not the purpose of this volume to go into detail regarding engineering design, it bears stating that my current engineering design methodology follows the Canadian Masonry Design Code — CSA S304.1. There are several reasons for working with masonry codes rather than codes that deal with concrete building:

- Clay is a significant portion of the aggregate mix. Soil mixes containing particles smaller than 80 microns are not allowed in code-conforming concrete design. This is true in both the US and Canada.

- In masonry design, there is no prescriptively defined minimum compressive strength. For code-conforming concrete design, there is a minimum 28-day compressive strength required in prescriptive building codes by both American Concrete Institute (ACI) and Canadian Standards Association (CSA) design standards. In order to achieve these strengths a mere four weeks after mixing and placing, relatively high amounts of Portland cement are required, and this increases the carbon footprint and embodied energy of the material to the same levels as conventional ready-mix concrete. Masonry design codes allow for the compressive strength of a wall system to be designed in accordance with the actual load requirements of the intended building and the particular material assembly of masonry units and mortar proposed.

- Within masonry codes, there are lower minimum requirements for amount and spacing of reinforcing steel. For many reasons, the minimum cross-sectional area of reinforcing steel to cross-sectional area of a given masonry wall element is considerably less than the minimum cross-sectional area of reinforcing steel in a comparable concrete wall element.
- There are established existing masonry building codes that work in rammed earth's favor, rather than against it. One of the most complete current national rammed earth building codes, from New Zealand, is based on an engineered masonry design methodology.

Embodied Energy and Carbon

While this book is not aiming to be a comprehensive source of information on the embodied energy and carbon in rammed earth, I will attempt to give the reader tools to help evaluate their design mixes with respect to these important, but not commonly understood characteristics. Like the other volumes in the *Sustainable Building Essentials* series, as well as Chris Magwood's *Making Better Buildings*, I am using the Inventory of Carbon & Energy (ICE) as the source for data on embodied energy and carbon for the materials used in rammed earth construction. The ICE database gives "cradle to gate" data for the manufacture and transport of materials only, and does not account for operational energy or thermal characteristics of the buildings created using them. The ICE was developed by professors Geoff Hammond and Craig Jones from the Sustainable Energy Research Team (SERT) in the Department of Mechanical Engineering at the University of Bath, UK. The version used in this book is V2.0, last updated January 2011. The database is open for use by the public, and the boundary conditions of its definitions and calculations are available for review. The database is hosted at circularecology.com.

The table below is a summary of the embodied energy and carbon for the basic ingredients used in rammed earth mixes, as well as several complete mixes (both raw and stabilized rammed earth) and comparable structural materials, like concrete. I have also included numbers for reinforcing steel and some common insulation materials. The first column to the right of the material's name is the embodied energy in mega-Joules per kilogram, followed by the embodied carbon in kilograms of carbon dioxide per kilogram of material, and finally the embodied carbon in kilograms of carbon dioxide equivalent per kilogram of material. The embodied carbon in kg/CO_2 is a measure of CO_2 emissions generated during the lifespan of the product or material in question. The embodied carbon dioxide equivalent, in kg/CO_2e, is a measure of all the greenhouse gases generated, normalized to CO_2. For instance, if 1 kg of methane (CH_4) is emitted during production of a given material, that portion would be added in as approximately 28 kg/CO_2e.

The data in the table is not likely to be accurately representative of the wide variety of material sources currently available in North America, but it nonetheless represents one of the best-researched collections for building materials, and it is still very useful for meaningful comparisons within its scope. That is to say, while the final number for any given material may not be exactly correct for that material in your particular neighborhood, all of these numbers are based on a consistent methodology and are useful for comparing with each other.

For example, it is interesting to note that the relative increase in embodied *energy*, measured in mega-Joules per kilogram, going from raw rammed earth to stabilized rammed earth with 5% Portland cement (by mass) is 0.68 − 0.45 = 0.23, or just over 50%. At the same time, the relative embodied *carbon* increase, measured in kilograms of carbon equivalent, from the addition of this small amount of stabilizer is 0.061 − 0.024 = 0.037, or more than 160%. This is a useful "apples-to-apples" comparison.

Full life-cycle analysis (LCA) and general accounting of embodied energy and carbon is beyond the scope of this book, but it is a rapidly developing field that promises better-informed design comparisons for all building materials in the near future. Environmental Product Declarations (EPDs) are becoming mandatory for manufacturers to provide in order to meet tender and specification requirements, and this holds promise of greater transparency when sourcing materials. Recent legislation in Washington State and California requires EPDs to facilitate carbon accounting in order to meet climate change commitments.

That said, a building material like rammed earth, which is not a consistently manufactured product, may find itself on the outside looking in if this type of accounting cannot accommodate site-built construction types. There are many natural building practitioners, designers, and academics working on this. I encourage the interested reader to look up the Embodied Carbon Network, which is part of the Carbon Leadership Forum. In particular, review the work of the task force on renewable materials for more current and detailed information.

Relevant Research

Arrigoni, Alessandro Marocco. "How sustainable are natural construction materials? Stabilized rammed earth, hempcrete and other strategies to reduce the life cycle environmental impact of buildings." Doctoral thesis, Politecnico di Milano, 2017.

Arrigoni, Alessandro, et al. "Life cycle analysis of environmental impact vs. durability of stabilized rammed earth." *Construction and Building Materials* 142, 2017, pp. 128–136.

Table 2.1: ICE values for embodied energy and carbon in materials relevant to rammed earth building

Material	Embodied energy in MJ/kg	Embodied carbon dioxide in kgCO$_2$/kg	Embodied CO$_2$e (carbon dioxide equivalent) in kgCO$_2$e/kg	Comments
Rammed earth — site soil	0.45	0.023	0.024	Raw earth — no chemical stabilizer added
Stabilized rammed earth — 5% cement	0.68	0.060	0.061	5% Portland cement added (by mass)
Stabilized rammed earth — 8% cement	0.83	0.082	0.084	8% stabilizer — 6% Portland cement, 2% lime (by mass)
Sand	0.081	0.0048	0.0051	UK data — heavily influenced by fuel/transport costs
Aggregate — general	0.083	0.0048	0.0052	UK data — heavily influenced by fuel/transport costs
Ground limestone	0.62	0.032	—	Not based on a large sample size*
Regular Portland cement	5.50	0.93	0.95	Normal Portland cement — 94% clinker, 5% gypsum, 1% minor additional constituents
Cement with 6%–20% fly ash	5.28 (6%) to 4.51 (20%)	0.88 (6%) to 0.75 (20%)	0.89 (6%) to 0.76 (20%)	Regular Portland cement amended with fly ash from coal-fired electricity generation
Cement with 21%–35% fly ash	4.45 (21%) to 3.68 (35%)	0.74 (21%) to 0.61 (35%)	0.75 (21%) to 0.62 (35%)	Regular Portland cement amended with fly ash from coal-fired electricity generation
Lime	5.30	0.76	0.78	ICE researchers noted that embodied carbon was difficult to measure for lime**
Concrete	0.75	0.100	0.107	25 MPa concrete — common compressive strength threshold for reinforced structural concrete ~12% cement binder by mass
Bitumen	51	0.38 to 0.43	0.43 to 0.55	42 MJ/kg feedstock energy included. CO$_2$ emissions are difficult to determine, so a range is given
Glass Fiber Reinforced Plastic (GFRP)	100	8.1	—	1998 data from the Steel Construction Institute
Steel rebar	21.60	1.74	1.86	World average recycled content
Cellular glass	27	—	—	No CO$_2$ data available in the ICE database
Cellulose	0.94 to 3.3	—	—	No CO$_2$ data available in the ICE database
Cork	4	0.19	—	2003 data from "Ecohouse 2: A Design Guide," Roaf, Fuentes, and Thomas
Mineral wool	16.60	1.20	1.28	2003 data from LCA documents 8, Eco-Informa Press
Rock wool	16.80	1.05	1.12	
Expanded polystyrene	88.60	2.55	3.29	46.2 MJ/kg feedstock energy included
Extruded polystyrene	109.20	3.45	4.39	49.7 MJ/kg feedstock energy included
Polyurethane rigid foam	101.50	3.48	4.26	37.1 MJ/kg feedstock energy included

Source: Inventory of Carbon & Energy (ICE) Version 2.0.

*Crushed limestone is included because it is similar in chemical composition to limestone screenings, which have been used with some success as a part of a stabilizer substitute/reducer for Portland cement. Limestone screenings, a by-product of aggregate extraction and processing in limestone rich quarries is not the same as industrially ground limestone, purposely made into a fine powder, nor does it have the same embodied energy, as the process is less intense. Crushed limestone is commonly used as a base for stabilized rammed earth in western Australia, and may be a good option for places where it is available and the site sub-soils are not suitable for rammed earth.

**There is no clear agreement in the scientific community on how to account for carbonation that occurs during the operating life of a lime-based element like an exterior plaster. The laws of thermodynamics dictate that it cannot be carbon neutral.

Chapter 3

Building Science Notes

Design for Site and Occupancy

BUILDING SCIENCE has become a necessary and growing field of specialization in the construction industry. With increased insulation and airtightness levels becoming mandatory under new energy codes, we need to think carefully about our buildings' assemblies and how they will perform under all of the conditions we subject them to.

I generally think of buildings as being alive — although not quite as animated as the people and pets that they shelter. Perhaps they are more like plants: not able to move much, subject to sun, rain, wind, and needing to deal with extremes of hot/cold and drought/flood. Much like a plant, a building has little choice but to deal with the site that it is placed into; it can't re-situate itself if it finds that the sun is a bit more accessible to the south, and it may have to wait ten years to benefit from the shade of a young tree. The building needs to breathe, in the sense that stale air is exhausted and fresh air is brought in. The indoor humidity needs to stay at a level that keeps occupants and materials healthy — not too dry, but also not so wet that mold and mildew begin to develop. The indoor temperature needs to stay within an acceptable range — not cold enough to allow plumbing to freeze, and not so hot that occupants are unreasonably uncomfortable.

The building can make use of the sun via proper window location and the incorporation of solar appliances. Wind and the stack effect can be used for passive ventilation in some seasons, and also as a source of energy. Rain can be captured and stored for use in everything from gardening to drinking, depending on the level of filtration and treatment one is willing to incorporate.

A site resource that I rarely see fully engaged is the steady temperature of the soils below buildings. Recently, insulating basement spaces has become more common, which does make sense, especially for the depth of the foundation from grade down to frost level. And I certainly advocate going beyond that on the perimeter of the building — more insulation on the vertical surfaces and around footings is better than less. Porous insulation materials like mineral wool are able to act as both insulation and as part of sub-surface drainage systems. Short of employing active systems such as ground source heat pumps, it is possible to make use of the relatively cool (summer) and warm (winter) temperatures of the soil below the frost line. This is not a book about mechanical design, but I encourage the reader to consult passive solar design resources, and to consider earth-berm north walls, or at least a walk-out, if a basement is part of the design.

That said, there are many reasons to avoid basements altogether — not the least of which is the carbon content and embodied energy cost of large volumes of concrete. There are other ways to build foundations, but that is a subject for another book.

How the building is situated, the location and quality of glazing, the overall geometry, the thermal mass, the insulation, and the airtightness are all passive elements that have an enormous impact on the living situation for the building and its occupants. Your site may have great views to the north, but I suggest building a gazebo or incorporating outdoor spaces to appreciate that view rather than installing large north-facing windows that would be a net energy loss — at least in the climates that I work in.

There are four control layers that have become touchstones for building science practitioners to consider when looking at building envelope assemblies:

- the *water control layer* (or weather barrier)
- the *air control layer*
- the *vapor control layer*
- the *thermal control layer*

In cooling-dominated climates like Australia or very moderate climates like Rwanda, a rammed earth wall can provide excellent performance by acting as all four layers in one material. In heating-dominated climates like Canada and the northern portion of the continental US, there needs to be a layer of insulation included to achieve a completely effective envelope. If constructed as a solid wall, rammed earth is simply too conductive to economically keep an interior space above freezing during several months of sub-zero weather.

However, incorporating an internal layer of insulation means that a rammed earth wall can act as *all four layers* with very high performance, even in heating-dominated climates. That said, we know that the walls aren't the only part of the building envelope. There are openings in the walls for doors and windows, and there are connections to be made with foundations, other walls, and roof systems. All the connections between these elements need to be designed and executed well, otherwise our labor-intensive, well-insulated, high-thermal-mass rammed earth wall won't make much of a difference in the overall building's performance.

Let's look at each of the four layers in more detail, one at a time.

Water Control Layer

Also known as the *weather barrier,* the water control layer is the outermost surface and/or assembly that has to deal with sun, wind, rain, snow, ice, and everything else the external climate throws at it. Here, the issue of durability raises its head. Exposed raw rammed earth can be compared to an earthen plaster finish. It will erode under driven rain and high wind, and will need maintenance over time; however, with a proper roof overhang and protection from backsplash and snow melt at the base of the wall, a single-story raw rammed earth wall can work in our climate.

The question of aesthetics needs to be addressed here. Many people are drawn to rammed earth for the way that it looks and feels, and they are loathe to cover it up. But with exterior walls, it is a given that they will erode. One solution is the application of a plaster layer — knowing

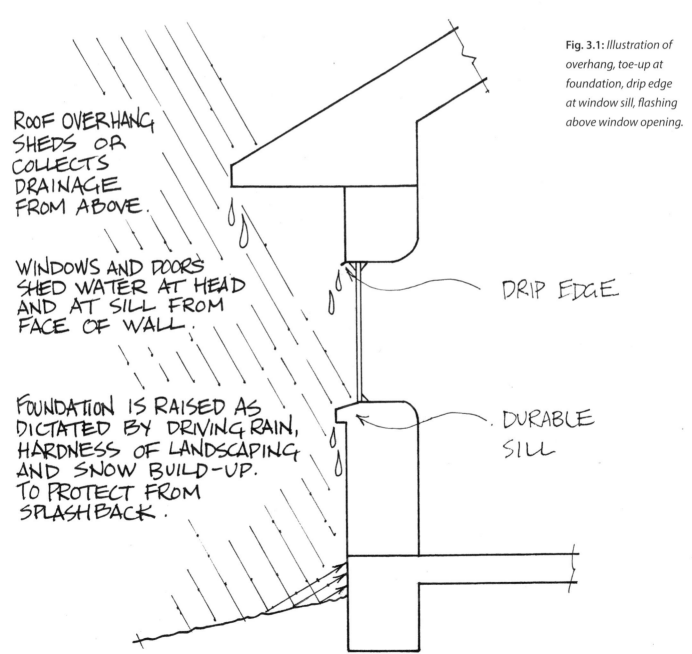

Fig. 3.1: *Illustration of overhang, toe-up at foundation, drip edge at window sill, flashing above window opening.*

that it will need to be renewed on a regular schedule. But, perhaps we are better served by using a different exterior material for the weather-control layer. If raw rammed earth is what you are aiming to build with, I recommend exterior insulation and a vented rainscreen to protect the whole assembly.

That said, practitioners in Europe have made considerable progress with raw rammed earth exteriors by incorporating various textures to reduce erosion by allowing water to run down the surface and by purposely adding thickness to the wall to provide a sacrificial layer. The idea of designing with extra material to account for erosion is not unlike the additional cover depth required for reinforcement in concrete that is exposed to high saline conditions, or the char-layer depth that timber engineers add to their structures in areas potentially affected by fire. From this broad engineering perspective, raw rammed earth can be designed as an effective weather-control layer with very interesting aesthetics. The interested reader is encouraged to look at the work of Martin Rauch in Austria, Switzerland, and Germany.

Stabilized rammed earth has the durability necessary to serve as an exterior finish, even in extreme climates where the walls can be expected to experience multiple wet/dry, freeze/thaw cycles annually. A multiple layer build-up (the rammed earth layers are called *wythes* in masonry terminology), with exterior and interior stabilized rammed earth wythes and an internal insulation layer will act as all four building science control layers for our ideal envelope.

Even in the jurisdictions that have the most demanding regulators and inspectors, an external stabilized rammed earth wythe can be considered equivalent to a veneer layer of masonry. In the absence of climate-specific durability testing, this exterior layer may not always be considered structurally load-bearing by plan examiners. That said, in less exacting jurisdictions, it may be taken as structurally "active." I find that it is good practice to consider the external layer a veneer; this promotes a structural separation between the interior and exterior, and discourages the inadvertent creation of thermal bridges across the wall assembly.

Another set of issues comes with insects, small animals, and vegetation. A solid rammed earth wall is an excellent barrier to insects and small animals — provided connections to windows and doors are constructed properly. Vegetation such as vines could be a long-term problem on a raw rammed earth wall, but it is extremely unlikely that there would ever be enough moisture deep in the wall to encourage significant root penetration.

A vented rainscreen is more vulnerable to insects and small animals and should have proper detailing at both the base and the top of the wall to minimize access for creatures looking to nest inside your building assembly.

Air Control Layer

Also known as the *air barrier,* this is an increasingly important aspect of high-performance construction in northern climates. Conditioned air — air that is at a more desirable temperature and humidity than exterior air — is at a premium inside our building envelope. If the conditioned air is allowed to leak through the building's skin, we are losing energy and potentially causing damage. In winter, we keep our indoor environment at a higher temperature and higher relative humidity than the outdoors. As this warm, moist air passes through the building envelope, it cools down — usually to the point where the relative humidity is more than 100%. For any given humidity level, there is a temperature where condensation will begin to occur — this is the *dew point.* Generally speaking, liquid water condensate forming out of air at atmospheric pressure requires an opaque surface to form on, such as the back side of sheathing in a framed cavity wall. So, we may feel that we are safe from the threat of this sort of damage with our monolithic rammed earth walls. Well, yes and no. While the rammed earth itself is largely immune to air penetration at any meaningful depth, the walls have openings and connections at foundations, floors, and roofs — as well as penetrations for utilities, exhaust, and makeup air ventilation. Detailing these locations properly is critical for the overall performance and long-term durability of the entire building assembly.

Air is a gas, and air movement is driven by differences in pressure, moving from high to low pressure. When we create an enclosed volume, we also create a pressure differential between the top and bottom of our building. This is known as the *stack effect,* and it is a direct consequence of the buoyancy of warm air. In the heating season, the outdoor air is cooler than indoor air, which causes a higher pressure region indoors at the top of the conditioned space. At some elevation below this, the indoor air pressure is equal to the outdoor air pressure. The location of the neutral pressure axis depends on wind conditions, building geometry, and airtightness, and it will vary over the course of the day. Figure 3.2 is a graphical representation of the air pressure distribution across a sample building envelope in the heating season.

In the cooling season, the outdoor air is warmer than the indoor air, and the pressure regime reverses across our building envelope. This is due to the corollary of our earlier statement — cool air is denser than warm air and will sink. The overall effect is usually less than in the heating season, as the temperature differential across the envelope in summer isn't nearly as high as it is in winter. Figure 3.3 is a graphical representation of the air pressure distribution across the same building envelope in the cooling season.

In the shoulder seasons, these air pressure distributions will switch back and forth diurnally, and sometimes over the course of a few hours. This will often also happen in cooling season when the temperature can drop significantly overnight. These annual (and daily) variations in pressure make designing passive ventilation systems very challenging in northern climates.

Ventilation is necessary in airtight assemblies to provide adequate O_2 for occupants and any combustion appliances, to remove excess CO_2, and to filter out dust, allergens, and other particulate matter. Do not be fooled into believing that having a "breathable" building envelope means that fresh air will pass through your walls. Quite the opposite. A properly built "breathable" envelope allows moisture to travel through while serving as an air barrier.

It may be possible, and even advisable, to provide adequate fresh air using windows and stack effect during certain times of the year — but it is still necessary to design and implement active ventilation for a safe and healthy built environment.

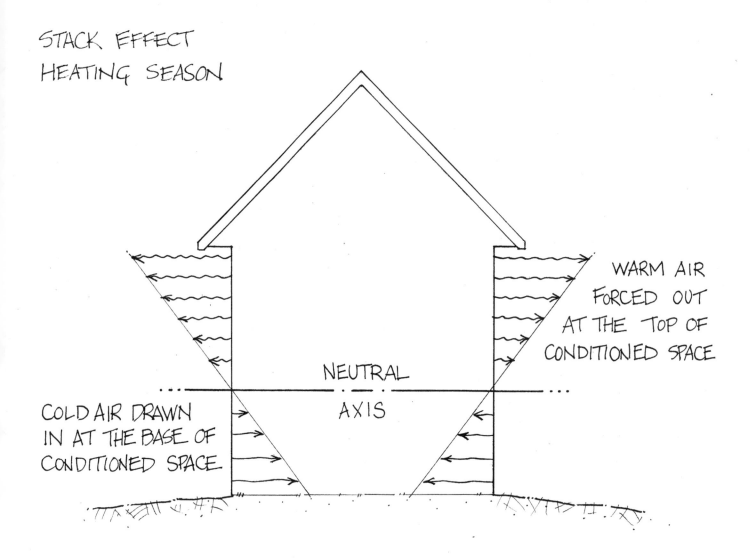

Fig. 3.2: *Stack effect in the heating season: outdoor air is colder than interior air.*

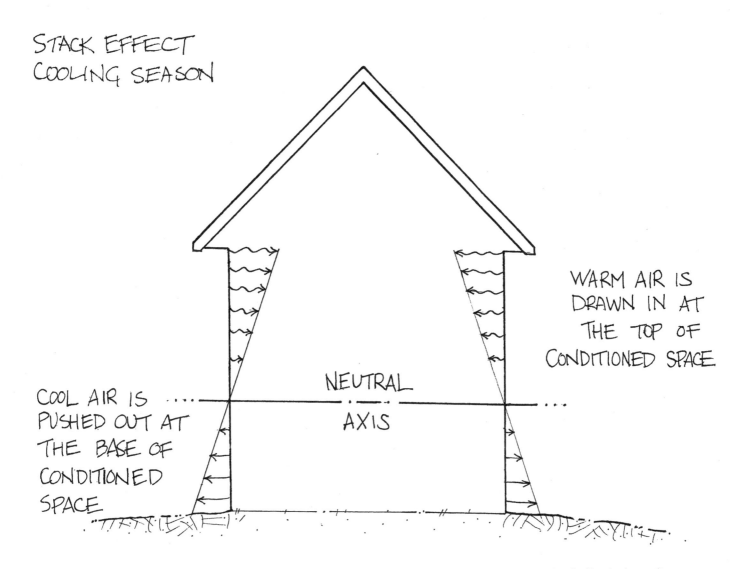

Fig. 3.3: *Stack effect in the cooling season: outdoor air is warmer than interior air.*

Vapor Control Layer

This is also known as the *vapor barrier*. Strictly speaking, the vapor control layer's only purpose is to limit the diffusion of water vapor through permeable components of the building envelope. This happens at the molecular level — the water vapor does not change phase, and air does not physically pass through the material. The movement of vapor by diffusion is driven by a moisture/humidity differential across the building assembly, otherwise known as a *vapor pressure differential*.

For many reasons, North American building codes have historically put an emphasis on the need for a vapor barrier rather than an air barrier; so much so that polyethylene sheet goods are often simply referred to as "vapor barrier" in building supply circles. This is not a wrong way to think, per se — but a well-detailed, continuous vapor barrier is also a very effective *air* barrier, and the performance that early energy-efficiency-conscious code officials were looking for was being carried out by the airtightness more than the vapor impermeability of wall sections. Where vapor barriers were not well detailed in northern climates, the damage in walls was far more often due to condensation due to air leakage than to moisture build-up due to diffusion.

While a continuous vapor barrier can be an effective air barrier, the opposite is not necessarily true. High-porosity materials such as some clay plasters and engineered membranes can be airtight while allowing vapor to pass through. This characteristic is often necessary in designing assemblies that need to be able to dry out over the course of an annual weather cycle. Most materials have some ability to store moisture during damp situations and then release it later during drier conditions. This is referred to as the *safe storage capacity* of the material — the amount of moisture it can contain at a given temperature before mold and mildew begin to form or other damage due to excess water is caused. Figure 3.4 is an illustration of the balance between wetting and drying cycles in an assembly, held in place by the safe storage capacity of the material.

While *vapor barrier* has become part of the vernacular, it is more accurate to think of materials in terms of their *permeability*, and to put them in three classes based on how they work as vapor retarders. In North America, the common unit of measurement for this is the *perm*.

- Class I — 0.1 perm or less (this qualifies as *vapor impermeable*, or as a true *vapor barrier*)
- Class II — 0.1 to 1.0 perms (*vapor semi-impermeable*)
- Class III — 1.0 to 10 perms (*vapor semi-permeable*)

Any material with a perm value above 10 is considered completely vapor permeable.

The moisture storage capacity of rammed earth is considerable, but it is difficult to quantify for any particular mix without extensive (and usually expensive) testing. A corollary of rammed earth's capacity to store moisture during periods of high humidity and then release it in drier conditions is the effect of ambient humidity buffering. This property is being studied within the context of passive cooling, although not yet extensively. What is clear at this time is that pore distribution within the rammed earth is the most important contributor to humidity buffering, and that raw rammed earth is more effective than stabilized rammed earth. This is because raw rammed earth has a higher porosity than stabilized. The contribution of clay in the raw rammed earth is not unimportant, but it is difficult to quantify in comparison with pore distribution.

Most studies agree that the effect of humidity buffering is limited to the first 50 mm (2") of

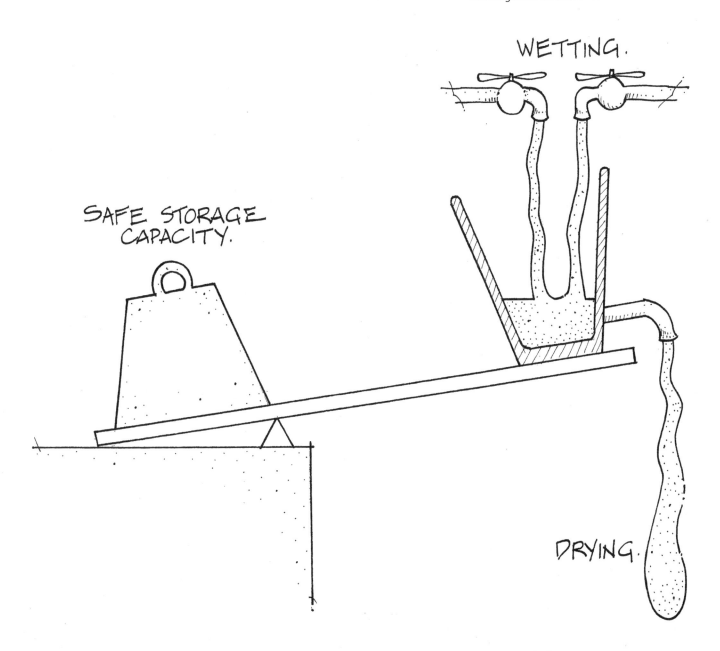

Fig. 3.4: *Balancing wetting and drying mechanisms.*

rammed earth. This suggests that rammed earth sections greater than 50 mm (2″) in thickness are an effective vapor barrier, and since the current method of pneumatic tamping does not lend itself to widths less than 150 mm (6″), a rammed earth wythe — whether raw or stabilized — can be considered a Class I vapor barrier.

Thermal Control Layer

In northern climates, the thermal control layer is commonly known as *insulation*. Rammed earth is not a particularly good insulator; in fact, it's a fairly good conductor, at least in comparison with wood, the most common structural material for residential construction in North America. So, we advocate for the addition of an insulation layer — whether that is inside the wall itself or external — in order to keep the heat (and cool) in. Notice that we do not recommend insulating from the inside of the building. Doing so would rob us of most of the benefits of thermal mass. Regardless of whether walls are made of raw or stabilized rammed earth, they are excellent thermal mass. Within an insulated building envelope, this mass can store heat during the day and release it overnight during the heating season. During summer, cool night air can be brought into the building without extreme drops in sensible temperature because of the thermal buffering effect of the mass of the walls. It takes a considerable amount of heat to raise the temperature of the mass, so it will feel cool during the heat of the day.

During shoulder seasons, the mass will easily keep the temperature of the interior steady without significant input. The caveats here are the need for a relatively airtight enclosure, proper solar siting, and an efficient ventilation system.

I encourage the interested reader to look at passive home heating books by Ed Mazria, Dan Chiras, and James Kachadorian — as well as the work of advocates of annualized geo-solar design. I'm sure there are other good sources for solar design out there as well, but these three authors' books contain a wealth of information about passive solar heating — and some good ideas for passive cooling as well.

It is possible to incorporate hydronic pipes into a rammed earth wall, although it is not especially practical to do so. The relatively high conductivity of the material itself means that it can radiate relative heat or cool into a volume at a rate that most humans can sense. This makes it possible to combine the thermal mass and radiative properties in a similar fashion, as has become common with concrete floor slabs. Vertical hydronic installation isn't highly recommended, as it adds considerable complexity to the forming and ramming steps, and it is of questionable value — that is to say, it is not low-hanging fruit, energy or comfort-wise. However, for an ambitious design, a vertical hydronic element could be part of a system that moved higher- or lower-temperature working fluid from one part of the building to another — potentially storing and releasing passive solar gain in an area that does not directly have solar exposure.

The type of insulation that is appropriate for a rammed earth wall system depends on its location in the wall assembly. If it is external to the rammed earth, it needs to be durable and able to support a weather barrier layer — either plaster or a rainscreen assembly. If it is internal within the rammed earth, it needs to be able to handle the pressures, dust, and vibration of tamping — but once the wall is built and the roof is on, it will not likely see any damage.

Rammed Earth's Dirty Secret

Most stabilized, insulated rammed earth built in North America has incorporated some kind of foam insulation — either polyisocyanurate, polystyrene, or polyurethane. This is for reasons of economy and regulatory compliance. The high insulation value per unit of thickness available with petroleum-based foam products is very seductive. To build a code-conforming (in terms of structure, durability, and insulation value) 2,400 mm (8′) tall wall in Canada that is a maximum of 450 mm (18″) thick would require two 150 mm (6″) stabilized rammed earth wythes on either side of 150 mm (6″) of rigid insulation. If the binder content is close to 10% by weight, and the insulation is any kind of petroleum-based foam, this wall system is no better in terms of embodied energy or carbon than a conventional 300 mm (12″) insulated concrete form wall with 75 mm (3″) of polystyrene on either side of a 150 mm (6″) 25 MPa concrete core.

In terms of operational energy, the effect of embedding the insulation between two layers of mass results in a significantly higher energy performance and interior comfort level than the conventional Insulated Concrete Form (ICF) building. This is definitely an advantage, but it is not necessarily unique to stabilized rammed earth.

I find this truth hard to swallow, but it is the truth nonetheless. I am not saying that rammed earth cannot be built sustainably in northern climates, but I do not believe that the majority of the North American rammed earth building that has been done to date can claim to be carbon neutral — which, in the light of global climate change, really needs to be a key criteria for building design.

I am not saying that rammed earth construction the way it is being done is worse than conventional concrete construction, but it is hard to say that it's anything other than *less bad*.

And *less bad* does not meet the long-term goals that our built environment needs to reach.

On the bright side, mycelium foam holds great promise for high insulation value per unit thickness, and cellulose fiberboard products are now readily available. Methods of removing lignin from wood to create very strong, highly insulative cellulose panels are being refined, and this technology is full of potential both in terms of structure and the building envelope.

Similarly, innovations in geo-polymers and calcium carbonate-producing enzymes are pointing to low-carbon stabilizer solutions that completely remove Portland cement from high-compressive-strength compressed earth mix designs. We are still some time away from these technologies entering the mainstream of masonry-style construction, but their development is very encouraging.

Relevant Research

Arrigoni, Alessandro, et al. "Reduction of rammed earth's hygroscopic performance under stabilization: An experimental investigation." *Building and Environment* 115, 2017, pp. 358–367.

Beckett, C. and D. Ciancio. "Effect of microstructure on heat transfer through compacted cement-stabilised soils." Australian Research Council, 2014.

Fix, Stuart and Russell Richman. "Viability of rammed earth building construction in cold climates." Ryerson University, May, 2009.

Hall, M. and D. Allinson. "Assessing the effects of soil grading on the moisture content-dependent thermal conductivity of stabilised rammed earth materials." *Applied Thermal Engineering* 29(4), 2009, pp. 740–747.

Hall, Matthew R. "Assessing the environmental performance of stabilised rammed earth walls using a climatic simulation chamber." *Building and Environment* 42, 2007, pp. 139–145.

Chapter 4
Materials

The earth in rammed earth is principally subsoil; that is, gravel, sand, silt, clay, water, and air, with no organic materials (no topsoil, roots, partially decomposed plant matter, fungi, etc.). From a builder's perspective, there are many questions to answer: Will material from the site be acceptable? If not, can it be amended by adding one or more mixes of other soils from a nearby source? What proportions of the main soil particle sizes are suitable? How much water is needed for optimum compaction? How much stabilizer will be needed to achieve the necessary freeze-thaw durability? The necessary compressive strength? Will mechanical reinforcing be necessary? If so, how much and what kind?

Soils

Soils are formed by the physical and chemical weathering of parent material — bedrock. Engineers divide soils into a descending scale of particle sizes: boulders, cobbles, gravel, sand, silt, and clay. For rammed earth, we are not concerned with boulders or cobbles. We look for the following particle sizes: gravel from 20 mm down to 2 mm diameter (for strength in the macro-soil skeleton); coarse to fine sands from 2 mm down to 0.06 mm (to fill larger voids between the gravel particles); silts from 0.06 mm to 0.002 mm (to further fill in voids); and clays from 0.002 mm and down (to act as our non-cementitious binder). If you have a geotechnical lab do a grain-size analysis on various samples, they may provide the coefficients of uniformity (C_u) and coefficient of variation (C_v) based on points taken from the grain-size distribution curves.

Geotechnical engineers look at grain-size distributions graphically on a semi-logarithmic scale, as shown in Figures 4.1 and 4.2. Sieve and hydrometer tests are used in the lab to determine the various proportions of particle sizes in a given sample. Figure 4.1 is a plot of a purely theoretical sample, with an even amount of soil passing each sieve size. Figure 4.2 has three different samples plotted on the same graph; these are actual samples from an aggregate supplier representing average characteristics of various materials they supply.

There are a number of soil classification systems that can be used to give a gradation label to any given sample. Table 4.1 is a list of typical sieve sizes used by geotechnical laboratories, along with the material classification for certain ranges of particle size. The Unified Soil Classification System (USCS) is most commonly used in North America. Table 4.2 lists the primary and secondary letter definitions used in the USCS. It is beyond the scope of this book to go into a detailed explanation of the

---

Do Your Subsoils Make the Grade?

The phrases *well-graded* and *poorly-graded* mean different things to different people. If we are talking about apples, a well-graded sample would contain fruit of the same size, shape, color, variety, ripeness, etc. If we are talking about gravel, a well-graded sample would contain rocks of the same mineralogy in a narrow range of diameters, with very little other material present. Speaking in geotechnical terms, a well-graded soil has a *broad variety of particle sizes,* and we want well-graded soil in our rammed earth mix. So, for our purposes, we should think along the lines of smoothly or evenly graded soils along a grain-size continuum.

methodology behind the USCS, but the table will show readers what the letters in a USCS description mean.

Of particular interest is the size boundary between coarse sand and fine gravel — approximately 2 mm diameter. These are pretty much the largest-size particles that can be held together by matric suction, or capillarity, when damp. Anyone who has built sand castles on the beach has a hands-on feel for how small the damp sand particles need to be to hold up without any binder. This mechanism is one of two primary physical reasons that raw rammed earth holds together so well. The other physical phenomenon is the binding quality of clay. The structure of clay particles leads to each one being electrically charged, which in turn means that water, itself a polar molecule, is adsorbed onto the surface of clays. While the thickness of the adsorbed water layer is an order of magnitude less than the thickness of the matric suction meniscus curvature, it is possible that this characteristic increases the range of humidity in which suction is effective. Your sand castle doesn't have a lot of resilience against either extreme wetting or drying — but a properly built raw rammed earth wall *with the right clay-containing aggregate mix* can be remarkably resilient, though it will still erode in direct driving rain or actual submersion in water, as in a flood. When building exposed rammed earth, the incorporation of stabilizers greatly increases the durability of the walls, particularly in climates where freeze-thaw cycles are common occurrences.

Table 4.1: Typical sieve sizes and soil particle classifications

Classification	Sieve Size (mm)
Gravels	37.5
	26.5
	19.0
	13.2
	9.5
Coarse sands	4.75
Sand	2.36
	1.18
	0.60
	0.425
Fine sands and silts	0.300
Silts	0.150
Clays	0.075

Table 4.2: Unified Soil Classification System (USCS), primary and secondary letters

Primary Letter	Secondary Letter
G: Gravel	W: Well graded
S: Sand	P: Poorly graded
M: Silt	M: With non-plastic fines
C: Clay	C: With plastic fines
O: Organic soil	L: Of low plasticity (wL < 50)
Pt: Peat	H: Of high plasticity (wL > 50)

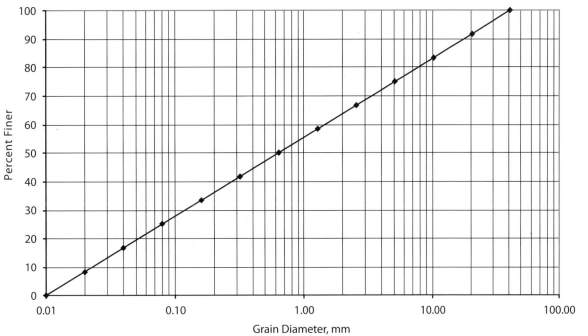

Fig. 4.1: *Theoretical evenly graded soil with particle diameter sizes ranging from 40 mm to 0.01 mm.*

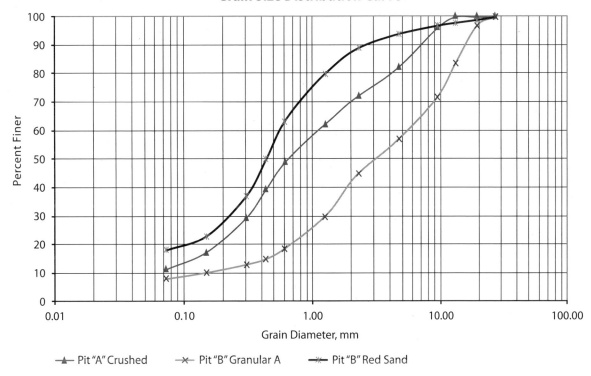

Fig. 4.2: *Actual grain-size distributions for several supplied aggregate products.*

Clay

Gravel, sand, and silt are formed by physical weathering, while clays are formed by both the physical *and* chemical weathering of parent bedrock. The primary structure of clay particles are crystalline sheets of silica and alumina. The sheets are combined and arranged in a variety of configurations, and are held together by hydrogen bonds or ionic bonds. Of particular note is the presence of water in the clay matrix — under normal atmospheric temperature and pressure conditions, it is virtually impossible to completely drive all of the water out of a clay sample because of the fundamental electrical charge inherent in the particles. This is as opposed to a sample of sand, for instance, in which the fundamental particles are electrically neutral; water will simply evaporate away from the sand if the vapor pressure or humidity is low enough.

Clay is the first — and oldest — binder used to make rammed earth, and we're going to "get into the weeds" a bit about it in the next few pages. If you don't want to go into great detail about the fascinating (to me and other geotechnical engineers and soil scientists, anyway) world of clay — you don't need to. That said, you do need to use clay in your raw rammed earth mix. Generally speaking, you can use most any clay, unless, of course, you can't — because the clay in your area is not suitable. You can find out if your area's clay is suitable for building by talking to farmers, potters, and plasterers who work with local natural materials. The clays to avoid are the *expansive* ones and the *sensitive* ones.

Expansive clays are those that swell excessively when fully saturated — and correspondingly shrink when they get drier. The clays in the Red River Valley in Manitoba, for instance, can vary as much as 300% in volume from wet to dry. These are *montmorillonite* clays, and they are not suitable for use in raw rammed earth (or pottery or natural plasters, for that matter). In dry years on the prairies of southern Manitoba, I have seen cracks in the soil in a regular pattern about 300 mm (12") apart that go down nearly a meter and are more than 5 cm (2") wide at the top. I have spoken to colleagues who report similar cracking in the clays near London, Ontario. Without serious fiber and cementitious stabilizer, anything built using these clays will tear itself apart with a few wet-dry cycles. It is not worth the risk to use these clays for building.

Sensitive clays are those that formed in a marine (saltwater) environment but are now in freshwater locations. The presence of water in clays allows for a great variety of ionic solutions, and the resulting chemistry of the pore water in any given clay can have a tremendous effect on the behavior of the overall material. The pore water chemistry of sensitive clays changes very slowly over time as the saltwater solution is flushed out and replaced with freshwater solution. This change in chemistry can lead to slope instabilities, and in extreme cases, sudden and catastrophic liquefaction. There is an infamous case of liquefaction that was caught on film in Norway in 1973. Interested readers are encouraged to search for the "quick clay landslide at Rissa." Here in Canada, there are sensitive clays along the St. Lawrence river in Quebec.

Happily, most clays are not this extreme. The chances are good that you will be able to use the clay from your area — but ask around first to see whether it's worth your time and effort, or whether importing clay from a known source is more practical.

A detailed explanation of the chemistry of clays is beyond the scope of this volume, but I do want to impress on the reader just how complicated and complex clay soils are. Because we have been looking at grain-size distributions based on particle diameters, we should try to get a feel for the size of clay particles. Figure 4.3 shows molecular models of (A) a silica tetrahedron and (B) an alumina octahedron. Figure 4.4

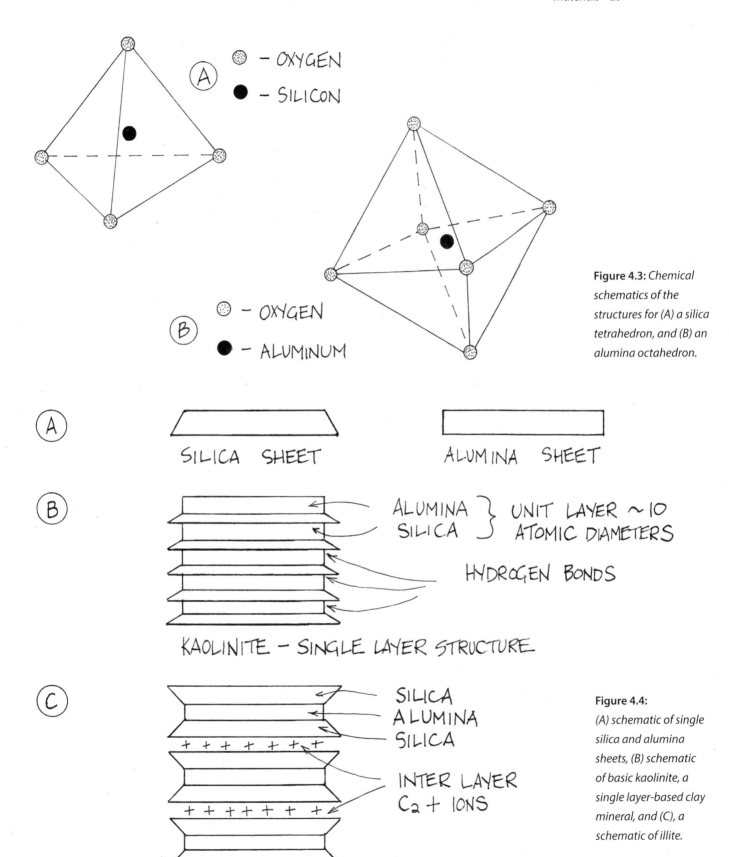

Figure 4.3: Chemical schematics of the structures for (A) a silica tetrahedron, and (B) an alumina octahedron.

Figure 4.4: (A) schematic of single silica and alumina sheets, (B) schematic of basic kaolinite, a single layer-based clay mineral, and (C), a schematic of illite.

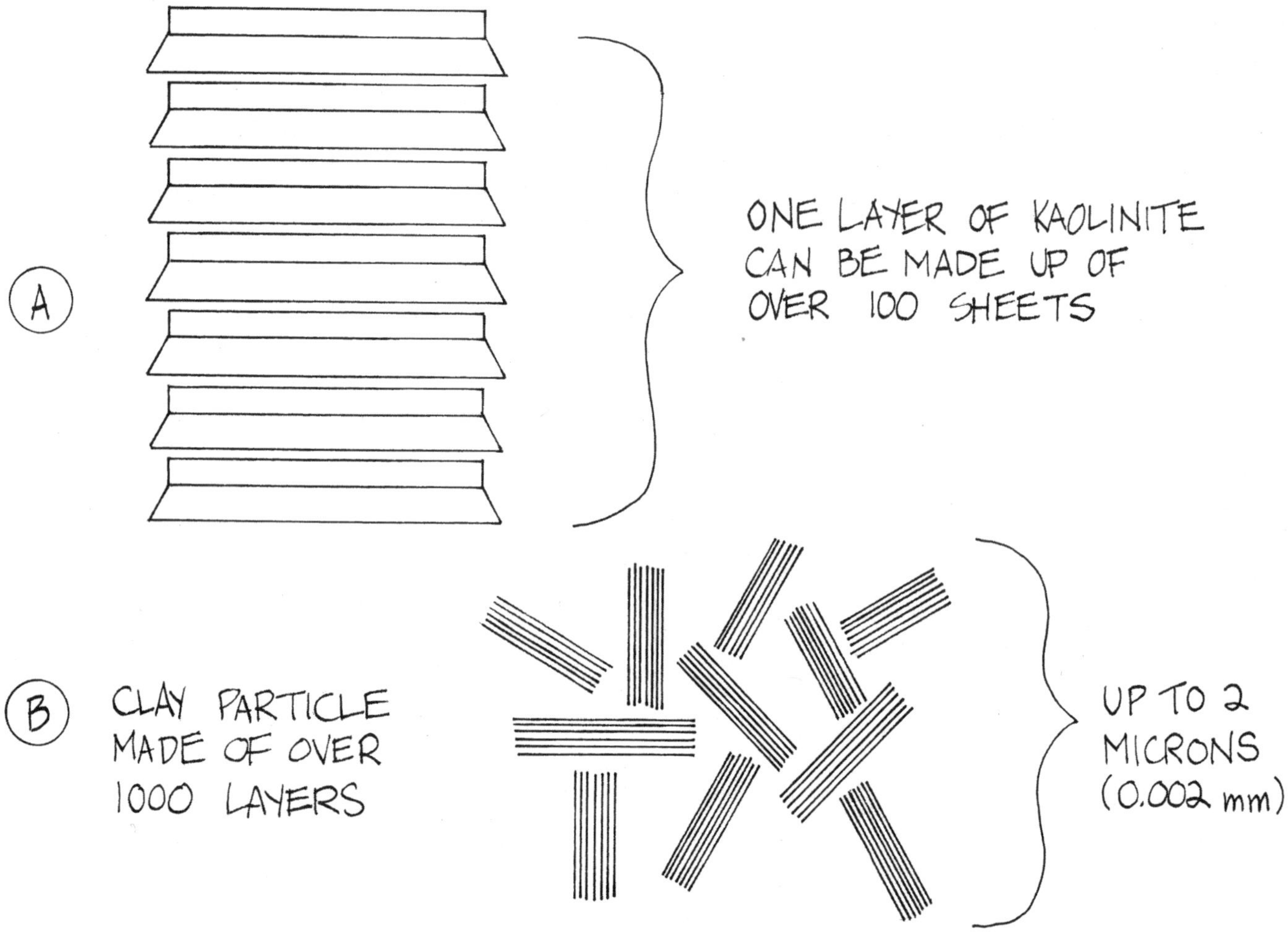

Fig. 4.5:
(A) a stack of single sheets forming a layer, and (B) an assemblage of layers forming a clay particle.

ONE LAYER OF KAOLINITE CAN BE MADE UP OF OVER 100 SHEETS

CLAY PARTICLE MADE OF OVER 1000 LAYERS

UP TO 2 MICRONS (0.002 mm)

shows a schematic of these particles and their basic arrangement in some typical clays. Figure 4.5 shows these particles in groups at scale.

For our purposes, we will look to the shrinkage characteristics and the plasticity of clays as the primary metrics for describing their suitability in our rammed earth mixes. Shrinkage is exactly what it sounds like — the amount a sample reduces in volume with loss in moisture content, usually expressed as a percentage. This is generally an easier property than plasticity for builders to consistently measure on their own, although there are test methods for plasticity that can be done without access to specialized laboratory equipment. Methods for both are described later in this chapter.

Plasticity is defined as *the ability of a clay soil to undergo unrecoverable deformation at constant volume without cracking or crumbling.* Plasticity is dependent on water content, and the plastic and liquid limits of a given clay are the water contents of the sample at the upper and lower ends of the range of plastic behavior. At the plastic limit, water content is relatively low, and the sample will tend to fracture rather than flex without cracking. At the liquid limit, water content is relatively high, and the sample will tend to flow rather than hold together in a cohesive mass. The plastic and liquid limits are also known as the *Atterberg limits,* named after the common test procedures to measure these characteristics. The Atterberg limits tests quantify a water-content range that is known as the *plasticity index,* which is the arithmetic difference between the liquid and plastic limits. The Atterberg limits are somewhat arbitrary, but they are determined using standard test methods. Different practitioners may arrive at different plastic and liquid limits for the same clay, but experienced lab technicians will be consistent within their own work. This is the classic "precision versus accuracy" issue. In your material selection process, it is better to have precision so that you can fairly judge between different mixes, rather than trying for absolute accuracy for the liquid and plastic limits. The numbers given in a laboratory test summary for liquid and plastic limits may not be the main criteria for your decision on whether any given clay is suitable for your mix. You may decide that it is not worth paying to have this testing done for your soils, but if the data already exists, it is well worth taking the information into consideration. Generally speaking, avoid clays with a high plasticity index, as they tend to undergo a high volume change with changes in moisture content.

Other metrics related to plasticity of clay soils are the *activity ratio,* which is the ratio of the plasticity index to the percentage of clay size particles in the soil, and the *shrinkage limit,* which is the water content of the soil when it reaches its lowest volume during the process of drying out.

Working with Clays

It can be quite difficult to achieve an even distribution of clay particles throughout a batch of earth taken directly from a site. Additionally, clays have a very wide range of characteristics that affect their behavior. Recall the schematics of the basic particle structures of clays shown in Figure 4.4. Kaolinite is the simplest, single layer-based clay around. It's the one that potters favor. It is consistent to work with, to fire, etc. The double-layer clays like illite, and its rowdy cousins montmorillonite and bentonite, are notoriously difficult to work with and do not behave well when fired. Bentonite is most usefully employed in the sealing of well casings and lining landfills; it is installed "dry" and then allowed to swell with exposure to groundwater.

There are entire classes of artistic pottery based on these highly expansive clays — Japanese Raku is an example. The main reason that these are considered artistic styles of pottery is because it is incredibly difficult to throw or turn and then fire a useful container with these double-layer clays. The difference in volume between wet and dry is just too great, and there are some clays with impurities in their pore water that make firing downright dangerous. High sulfur content in clay can be especially hazardous for potters (as well as brick makers). All of this is to say that simply calling your soil "clay" doesn't adequately or accurately describe your material for the purposes of earthen construction.

This variation in characteristics of any given clay is another reason to consider using a stabilizer in your rammed earth mix. Even high-plasticity clays with enormous shrink/swell potential have been successfully remediated using lime-cement mixes. Road builders, particularly in the UK, have been able to create durable road bases using in-situ clay soils amended with lime-cement blends. That said, if you are considering a raw rammed earth mix, you need to know the characteristics of your clay very well, and you must be able to ensure that it is evenly and consistently distributed throughout all of your batches.

There is a large range of recommended particle-size proportions for rammed earth mixes in the literature, but I will focus on two sources for our purposes: *Rammed Earth: Design and Construction Guidelines* by Pete Walker, et al.;

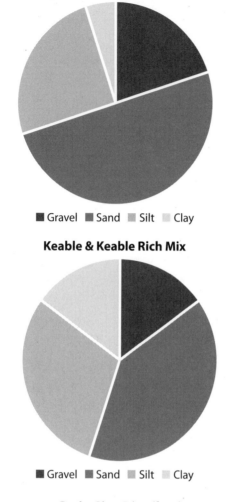

Fig. 4.6a: Pie chart showing relative soil particle size proportions of Keable & Keable Lean mix.

Fig. 4.6b: Pie chart showing relative soil particle size proportions of Keable & Keable Rich mix.

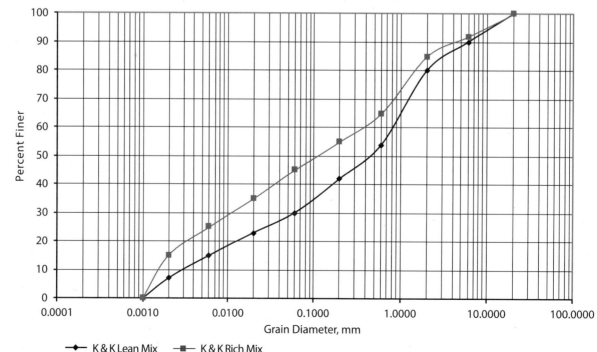

Fig. 4.6c: Grain-size distribution curve of Keable & Keable Rich & Lean mixes.

and *Rammed Earth Structures: A Code of Practice* by Julian Keable and Rowland Keable. The particle-size ranges given in these books effectively cover the practical mixes that a builder would use and are based on a large number of projects in the UK, Europe, and Africa.

The range of particle sizes recommended by Keable and Keable is 50%–70% sand and gravel, 15%–30% silt, and 5%–15% clay — plus the condition of a 80 to 120 mm break-off length in the "Roll Test," which is described below. Figure 4.6a–c shows pie charts and a grain-size distribution curve for two mixes based on the outer limits of the clay portion of this recommended mix. The lean mix represents the lowest limit of clay allowable, while the rich mix represents the highest limit of clay allowable.

The range of particle sizes recommended by Walker, et al. (by mass) is 45%–80% sand and gravel, 10%–30% silt, and 5%–20% clay; the plasticity index is 2 to 30 (liquid limit < 45); linear shrinkage <= 5%; soluble salt content < 2%; and organic matter content < 2%. Figure 4.7a–c shows pie charts and a grain-size distribution curve for two mixes based on the outer limits of the clay portion of this recommended mix. The lean mix represents the lowest limit of clay allowable, while the rich mix represents the highest limit of clay allowable.

These ranges in both cases are given for raw earth designs, with clay as the only active binder apart from matric suction. However, staying within these proportions will probably not work equally well for stabilized rammed earth mixes.

Generally speaking, adding pozzolanic binders allows the use of less-evenly graded mixes and adds tolerance for the presence of poor-quality clay — or an insufficient amount of clay. It is usually beneficial to have less clay in a stabilized rammed earth mix. Higher amounts of silt, however, can be tolerated in a stabilized mix, and can even enhance the fineness of surface texture. A mix design example is given later in this chapter that describes the procedure for selecting an aggregate mix for a stabilized rammed earth project. An aggregate mix is a combination of different single products blended in an appropriate ratio for the given design.

While there is historic precedent for using other minimally processed binders in rammed earth, like bitumen or casein (dairy protein), in this book we will only consider pozzolanic

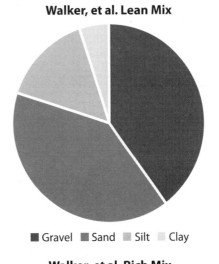

Fig. 4.7a: *Pie chart showing relative soil particle size proportions of Walker, et al. Lean mix.*

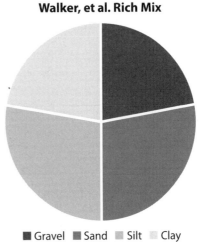

Fig. 4.7b: *Pie chart showing relative soil particle size proportions of Walker, et al. Rich mix.*

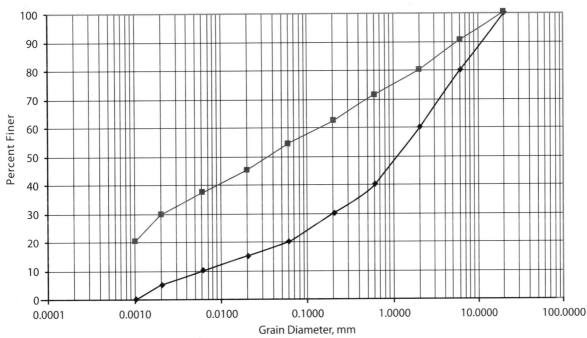

Fig. 4.7c: *Grain-size distribution curve of Walker, et al. Rich & Lean mixes.*

binders for our stabilized mixes. It is also possible to mechanically stabilize rammed earth using fibers like straw or polypropylene, but since the assumption here is that we are working toward an exposed wall with a certain aesthetic, we won't consider fiber reinforcement. The interested reader is encouraged to research earthbag building to see how the combination of a high-tensile-strength fabric and a compression-resistant soil mix can be used to form walls, domes, and vaults. Since these structures all require plastering or some other kind of cover, we are not going to go into detail about them here. There is a very good book on this subject in the Sustainable Building Essentials series, *Essential Earthbag Construction* by Kelly Hart.

Other Binders and Additives: Stabilized Rammed Earth
Portland Cement

Portland cement is the product of crushed limestone fired at high temperatures (at or above 1,300°C, or 2,370°F), along with certain amounts of other secondary materials. After firing, it is ground into a powder. Bruce King has authored an excellent book, *Making Better Concrete,* which covers the history of pozzolans, including Portland cement, and he makes a strong argument for using fly ash in concrete mixes at a relatively high proportion. Fly ash is an excellent pozzolan; it is a by-product of burning coal for steam power generation. A similar pozzolan is made from the residue of steel manufacturing, known as ground granulated blast-furnace slag. Its inclusion in concrete mixes tends to increase workability and set time, while reducing shrinkage. If you have access to fly ash, I recommend experimenting with off-setting Portland cement in your mixes between 25% and 50% (i.e. a range of binder blends of 3:1 Portland to fly ash all the way to 1:1). I say *if* you have access, because in many parts of Canada there are no large producers of fly ash due to the virtual elimination of coal-fired energy plants and the severe reduction of local steel manufacturing. Large concrete supply businesses might be able to supply you with fly ash, but they may require a large minimum volume per order, which could put it out of reach for small projects.

Fernando Martirena, a materials scientist in Cuba, has been working with colleagues from Switzerland and India on a product they call Limestone Calcined Clay Cement, or LC3. An excellent summary of his work is presented in *The New Carbon Architecture,* by Bruce King. For our purposes, LC3 can be substituted 1:1 for Portland cement, representing a considerable reduction in carbon emissions and embodied energy in the binder portion of our rammed earth mix. I do not have access to precise numbers to quantify the reductions, as this is a relatively new product at the time of writing. It is so new, it may not be available in your area yet.

One product that has recently become available in our area is GUL (ground-up limestone) Portland cement. This product incorporates 10% crushed limestone to offset the total amount of Portland cement in the blend. Producers claim that it can be used in mixes exactly as "pure" Portland cement would. This agrees with my experience: in several projects I have worked on, when the aggregate blend was made up partly of crushed limestone that included fines, higher compressive strengths were observed in test samples, alleviating the need for an increased amount of binder.

There is considerable work being done on cementitious materials all over the world. It is a rapidly evolving field, and any summary given in this book is likely to be out of date shortly after publication. It is my hope that the general principles included here will be useful to builders regardless of the precise product being considered as a stabilizer.

Hydrated Lime

Although it is similar to Portland cement, hydrated lime is produced in a process that occurs at lower temperatures. Crushed limestone is fired at a relatively high temperature (900°C, or 1,650°F) to form CaO, or quicklime. The quicklime is then mixed with water to form $CaOH_2$, known as *slaked*, or hydrated lime. The chemical reaction that occurs when water is added to quicklime is highly *exothermic*, meaning a lot of heat is generated and released in a short (and sometimes violent) period of time.

There are several reasons that lime is considered to have a lower carbon footprint than Portland cement. First of all, since the parent material is fired at a lower temperature, less energy is required for the initial part of the manufacturing process. Secondly, lime-based building materials will continue to undergo carbonation as long as they are exposed to CO_2 at the appropriate temperature and vapor pressure ranges. That is, the lime will take CO_2 molecules out of the air and form $CaCO_3$ (limestone) over a long period of time. As mentioned in the ICE-based table from Chapter 2, there is not a lot of quantitative information available to give hard numbers for a carbon footprint comparison of lime with Portland cement. I am sure that these numbers will become available as the importance of carbon accounting in building grows. Environmental product declarations (EPDs) hold the promise of being a source of data for an apples-to-apples comparison of these (and other) binders.

Calcined Clay

This is also the product of a high-temperature process, but it is generally considered a by-product, as it is not the principal material being refined. Ground kaolin clay is used as a dry lubricant in the high-temperature processing of materials in drum kilns. One such process starts with ground recycled glass, which is heated in rotary kilns to make a porous, lightweight aggregate. The clay lubricant is also subjected to the heat, and is called "calcined" clay after the process. An example manufactured in central Ontario is sold under the trade name "Metapor" by the Poraver company. Calcined clays have long been known to combine with lime to form pozzolans — i.e. Roman cement. Again, I refer the interested reader to *Making Better Concrete*, by Bruce King.

Oxides

These are chemically stable pigments added to the mix either in powdered or liquid form. Most are iron oxides — essentially, what we commonly refer to as *rust*. Not all oxides are red, but the colors available are limited by basic chemistry.

Oxides are added with the rest of the binder, and they should be blended as thoroughly as possible ahead of their addition to the overall mix. The maximum amount of oxide to add to a mix is around 10% of the total binder by weight — more than this will not increase the amount of color or add to its "brightness." In terms of increment, 0.5% by weight of total binder will change the visual nature of the mix. Minimum amount depends on the color of the basic material itself as well as the color imparted by the other parts of the binder mix. Common Portland cement is gray, and this definitely has an effect on the final color of the total mix. Portland cement is available in white, but usually at a 15% to 20% price premium. Lime and calcined clay both come in white exclusively, so they are less likely to affect the final color of any given mix.

Sealers

These are typically emulsions of silica — very fine silica particles suspended in water via a proprietary surfactant. Sealers are added to stabilized rammed earth during mixing to reduce efflorescence and the permeability of the material to liquid water. They greatly increase the mix's resistance to spalling due to freeze-thaw cycling by limiting the long-term moisture exchange to the vapor phase.

Tech-Dry, an Australian company, makes a product specifically for stabilized rammed earth called Plasticure, which is added at a rate of 0.5 liter (2⅛ cup) per 1,000 kg (2,200 lb) of dry ingredients. Plasticure has been used by many builders worldwide working with cement-stabilized rammed earth as an internal sealer and efflorescence reducer. It is a tried-and-true product, but its availability may be limited in parts of North America.

There are various *silane* and *siloxane* sealers available in the North American market, some of which can be added to the mix directly as an internal sealer. Many are intended to be applied externally after the forms are stripped. Externally applied sealers may be prone to degradation from UV and erosion, and they will usually have to be reapplied periodically.

Stabilized Rammed Earth Mix Design Example
Blending a Product from Aggregate Suppliers

Take a look back at Figure 4.2. There are three curves presented, each based on a sample taken by one of my clients from two aggregate suppliers. Sample 1 is from Pit "A," and is classified by the supplier as a Crushed Sand and Gravel. The supplier notes that approximately 55% of this mix is made up of crushed particles — i.e. particles that have been mechanically broken up, including all of the fines created by this process. Sample 2, from Pit "B" is classified by the supplier as Granular "A" with approximately 52% of the mix being made up of crushed particles. Sample 3 is also from Pit "B," and is classified by the supplier as a Red Sand, with no indication of any crushed particle content.

For reasons involving transportation distances and supply assurances, the product from Pit "B" was determined to be the best choice for the client. However, neither the Granular "A" nor the Red Sand product had the right grading for a good rammed earth mix. Fortunately, the supplier was willing to blend their product at a requested proportion, since a relatively large volume of material was being ordered.

To determine what proportion to order, the client made samples using 2:1 and 3:1 Granular "A" to Red Sand. Figure 4.8 shows the resulting grain-size distributions for these blends. At the same time, we tested different amounts of stabilizer to make sure we were achieving the necessary compressive strength for the design of the walls to be legitimate. Table 4.3 shows the test matrix for this set of tests.

The binder blend was made up of 2 parts Portland cement, 1 part slaked lime, and 1 part calcined clay (metakaolin), added in two concentrations: 7% and 10% — measured by weight of the total mix. Six cylinders of each mix were made, in order to test three at 28 days and the rest at 56 days. The samples were stored in a cool place, covered to avoid drying out too quickly. Because of the lime-calcined clay portion in the binder mix, strength is developed more slowly than with Portland cement as the only binder. (Details of compression strength testing are covered later in this chapter.)

In this case, the aggregate blend with a higher concentration of larger aggregate was stronger

Table 4.3: Test matrix of aggregate and stabilizer blends

	2:1 aggregate blend	3:1 aggregate blend
Stabilizer blend at 7%	6 cylinders	6 cylinders
Stabilizer blend at 10%	6 cylinders	6 cylinders

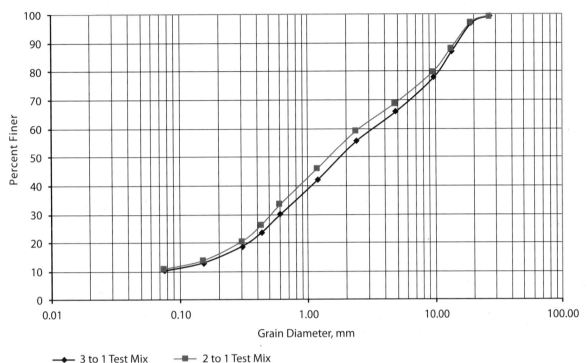

Fig. 4.8: Blended aggregate grain-size distribution. 2:1 and 3:1 mixes.

(i.e. the cylinders made with the 3:1 aggregate blend were stronger than those made with the 2:1 blend). And, as may be expected, the cylinders with the higher concentration of binder were also stronger.

Figure 4.9 shows in one graphic all the grain-size distribution curves we have considered. This can be thought of as representing the range of possible mix distributions. The bands on either side are guides showing the extremes of what can be roughly considered acceptable for a rammed earth mix. Successful raw rammed earth mixes will tend to have more clay content, while successful stabilized rammed earth mixes may have less fines (but more silt, depending on the desired finish texture). Regardless of what your grain-size distribution curves look like, it is always advisable to make and crush test cylinders, as well as larger-scale prisms and/or test walls before settling on a final mix recipe.

It should be noted that this example does not include tests done to determine color. Because color can be influenced as much by the inherent color of the soil blend as by the color of the binder mix, many more tests may be needed to identify a desired hue. For example, adding different oxide concentrations and possibly even different Portland cement color may require a large increase in the number of cylinders that need to be tested.

NOTE: Here and elsewhere, you will see measurements given in both volumes and weights. Ideally, everything would be measured by weight, which is far more consistent than volumetric measurement (ask any professional baker). But it's difficult to weigh the aggregate at the scale we're working with. So we end up mixing the two methods. At least we can encourage the binder mix + water/oxides/admixtures to be done as accurately as possible.

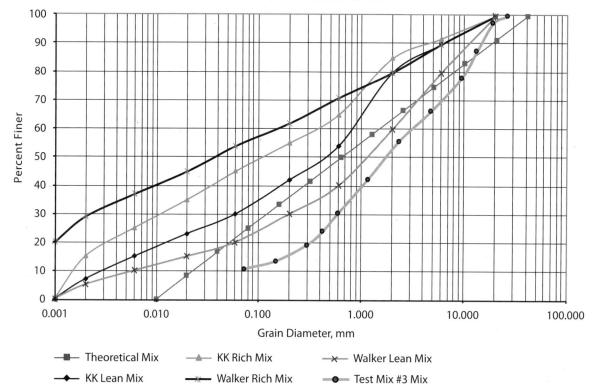

Fig. 4.9: Full range of grain-size distribution curves showing a range of possible rammed earth mixes.

Testing Procedures: Design and Construction Phases

This section includes a number of soil tests that have benefited builders for centuries, along with some procedures that are more state of the art. Many of the field tests are aimed at helping the raw rammed earth builder, but they can be useful for mix design and construction of stabilized rammed earth as well. The laboratory tests are important for engineering and regulatory purposes, but you can save time and lab costs by performing as many of the "hands-on" field tests as possible to narrow down the range of mixes being considered before engaging paid professionals.

Test walls are going to be your best tools for learning about mix design for raw or stabilized rammed earth, and they also offer insight into formwork techniques.

It bears stating that the mix selection process for *stabilized* rammed earth is quite a bit less sensitive to both the overall clay content and the type of clay present in the soil being considered. Many of the tests listed below are only necessary when designing a *raw* rammed earth mix. These are prefaced with a *Raw* designation. Tests that are not optional are designated with three stars: ***Raw*** or ***Stabilized***

Raw = The makeup of site materials is more important for raw rammed earth than for stabilized. Blending site clays into a mix is very difficult, but it can be done; however, it usually requires mechanical mixing equipment. If your site material is going to work on its own — and in an economical fashion — it will be because it has the clay already present. Amending with sand and gravel is possible (and preferable), but adding clay is not easy.

Stabilized = The consistency of aggregate materials is the more compelling component for stabilized rammed earth. Sourcing material from quarries that supply ready-mix factories is a good way to ensure you have the right gradation for stabilized rammed earth. If you are choosing stabilized over raw rammed earth, it is probably because your design requires higher strength. The strength of the mix is very highly impacted by the type and amount of binder, by the moisture content at time of ramming, by the delivery and tamping procedures, and by curing conditions.

Design Phase Testing
Field tests

These tests are carried out on site or at the source of potential soils being considered for construction. The primary purpose of these tests is to get a feel for the soil mixes being tested; the objective is to cost-effectively narrow down your choices before getting into compression cylinders and durability test walls, which are covered below, in the shop and laboratory test sections.

Raw Smell Test: Used to determine if there is organic matter in the soil. Organics lend a musty odor; "clean" soils are close to odorless. It is useful to dampen the soil if it is completely dry to the touch.

Raw Nibble/Taste Test: Used to determine quality of fines. It is essentially a texture test. Take a pinch of soil and rub it between your molars: sandy soils have coarse grit, which irritates the teeth; silty soils do not have an irritating texture against the teeth, although they are still gritty; clay soils are much smoother, feeling almost like a flour paste — they may stick to the tongue if the clay content is high enough.

Raw Ball Drop Test: This is a quick and dirty test to check if a given mix is at a suitable moisture content for ramming. While it is a "tried and true" method used for many centuries, it has not proven particularly accurate or repeatable across different practitioners or for different soil mixes. Nevertheless, it is an effective way to check a mix in the field in a timely manner, or to narrow in on the initial moisture content to minimize the number of samples you'll need for more accurate testing or laboratory tests, which can be expensive.

- Take a handful of the mix and pack it into a ball that fits in your palm. Hold the ball out at shoulder height and drop it onto a clean sheet of plywood or similar smooth, hard surface.

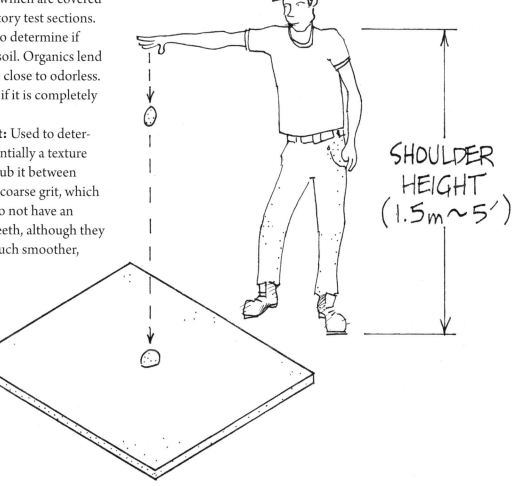

Fig. 4.10: *Drop Test: drop ball onto a hard, flat surface from shoulder height.*

- If you are unable to make a ball that holds together in your palm, the mix is too dry, or it doesn't have enough clay content, or both.
- If the ball stays together after dropping and doesn't crack appreciably, it is too wet for optimum compaction (and it likely has a large quantity of clay particles).
- If the ball shatters into many small pieces, it may be too dry for ramming and more water needs to be added, or it may not have enough clay content.
- If the ball breaks into only a few pieces, it has the right moisture content and contains the right amount of clay for ramming.

Variation — ***Stabilized***: Since most stabilized rammed earth mixes are drier than raw rammed earth mixes, the first half of this test becomes more important. If the mix forms a ball in your hand without falling apart, it is ready to ram. Generally speaking, a ball made of a good stabilized rammed earth mix at a moisture content ready for ramming is going to shatter into many small pieces if dropped from shoulder height; the same result would indicate a failed test for a raw mix.

Raw Slap Test: Used to determine the approximate proportions of silt and clay fines in a soil sample. Take a small handful of soil and wet it enough to make a cohesive ball about the size of a regular chicken's egg. Do not use more water than necessary to make the soil stick together — it should not stick to your hand as you are working it. As with the ball test, if you can't make the soil stay in a ball in the first place, the soil doesn't contain a significant amount of fines and is not suitable for raw rammed earth.

- Flatten the ball and hold it in the center of your palm. Slap the side of your hand against your other palm, or a convenient hard, solid object (don't injure yourself — if you do, you're doing it too hard). This action is meant to agitate the sample and bring the pore water to the

Fig. 4.11: Drop Test: close-up illustration of three basic results for the ball after dropping.

surface. The soil will appear shiny to varying degrees depending on the silt and clay content. If it does not appear shiny at all, there is not a significant amount of fines present.

- If the sample appears shiny in a few taps, say 5 to 10 at most, it is mostly fine sand and silt. You should be able to crush the sample in your fist, causing it to crumble.
- If it takes considerably more taps to make the sample take on a wet sheen, say 20 to 30, it has a moderate amount of clay in it. You should not be able to cause the sample to crumble; it will remain "plastic" when you squeeze it in your fist.
- If you cannot make the sample shiny with any amount of tapping, it has a very high clay content.

VARIATION 1 — TROWEL TEST: Instead of using your palm, place the sample on the flat of a masonry trowel and tap the trowel against a solid object. The same reactions as above apply.

VARIATION 2 — CUT TEST: Keep the sample in a ball shape and cut in half with a sharp knife. If the cut surface is shiny, it has a high silty clay content. If the cut surface is dull, it has a high clay content.

Raw Roll Test: Determines suitability of raw soil with respect to clay content/cohesive behavior, and gives an indication of optimum moisture content for ramming. This is best done after the ball and slap tests have helped narrow down the soil mixes being considered for further testing. Results from this test will be affected by the presence of pieces of gravel larger than 6 mm (¼") in diameter.

Take one of the "successful" balls from the earlier ball test and crush it to remove any lumps. Then slowly add water and work the ball with your hands until it is a smooth texture. The ball should be somewhere between a chicken's egg and a hardball in size. Set the ball on the ground or other hard surface. Take a 500 mm (20")

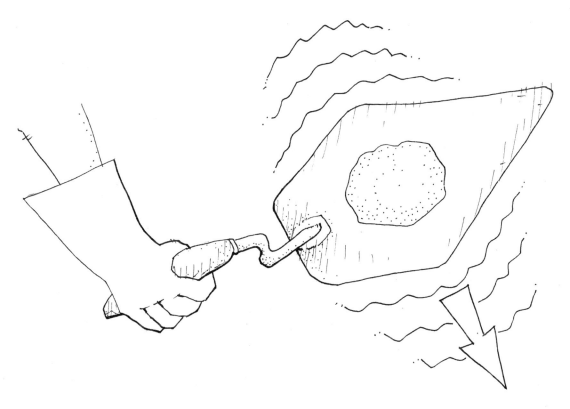

Fig. 4.12: *Slap Test: trowel variation.*

Fig. 4.13:
Step one of Roll test: determining moisture content.

length of 10M (#3) rebar and stand it vertically with one end resting on top of the ball of soil. Allow the rebar to sink under its own weight. Measure the depth that the bar penetrates the ball; ideally it will go in about 20 mm (¾″). If it doesn't go in that far, you can add more water and start again. If it goes in further, then you have more water than is good for compaction.

Now take the ball and form it into a cylinder 25 mm (1″) in diameter and 200 mm (8″) long by rolling it between your palms. Place the roll down on a table, perpendicular to the edge and push it along so that it extends out over open air. Slowly continue to push the roll out over the edge until it breaks off. Measure the length of the piece that breaks off. If the piece is less than 80 mm (3⅛″), then there is not enough clay in the mix to make a raw rammed earth structure. If the length of the piece is more than 120 mm (4¾″), then there is too much clay in the mix to make a raw rammed earth structure.

In very early sampling on site, this can all be done with the hands — no table or hard surface is necessary for initially determining if a soil source has the appropriate amount of clay to work with.

Raw Wash Test: In the most simple terms, this is determined by whether or not you need to wash your hands after conducting one of the preceding (like the roll or ball) tests. If you can rub your hands relatively clean as the soil dries,

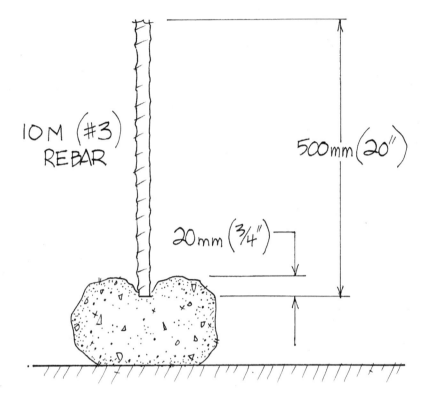

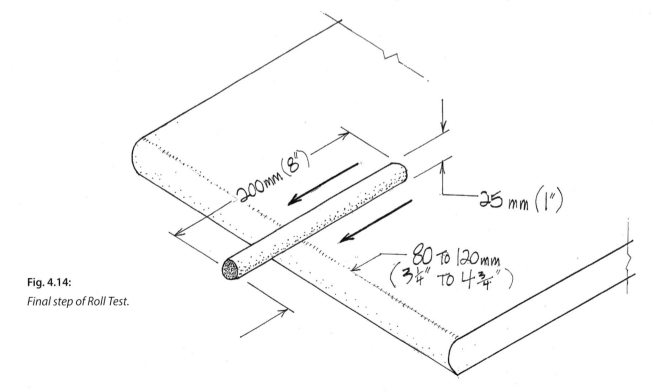

Fig. 4.14:
Final step of Roll Test.

it is higher in silt and sand. If you need to use water to get your hands clean to the eye, it is higher in clay. The texture of the soil against your skin is also useful to note; higher sand content will feel noticeably gritty, even with fine sands. Silts feel less gritty, but are not slippery when wetted. Clays, in particular those with a double-layer structure, can be extremely slippery and sticky when wetted.

Shop tests

These are tests that are carried out in a more controlled environment on mixes and materials that have been pre-selected based on on-site field testing. Again, the point of all of this is to get a feel for the material, to understand how moisture content at time of ramming affects the final density, shrinkage, and cracking characteristics of the mix, and how the addition of stabilizer, color, and permeability admixtures change the workability and finish of the material. It may not be easy to do, but testing in different ambient temperature and humidity situations can help you avoid issues that you wouldn't know about by only testing in a controlled environment like a lab. You want to have a thorough understanding of your mix and know the parameters that you can and cannot control before putting material into your full-scale forms and tamping it.

Raw Ribbon Test: This is a more refined version of the roll test, where the size of the roll is made consistent and multiple samples are tested to increase the representative accuracy of the results. We are looking for an indication of the suitability of the soil mix's clay content — and also honing in on the correct moisture content for ramming.

A jig is made on a smooth surface, such as a table top or workbench, with two 6-mm (¼") high strips set 20 mm (¾") apart. The soil sample is pressed into the space between the two strips and rolled with pin or smooth-sided bottle to make a consistent rectangular ribbon at least 300 mm (12") in length. To prevent the ribbon from sticking to the surface, it may be necessary to line the space with plastic or treat it with form oil or similar release agent.

At the edge of the work surface, create a curve with a radius of 10 mm (⅜"). Alternatively, you could use half of a 20 mm (¾") diameter dowel — this is to eliminate the effects of moving the ribbon out across a sharp edge.

As with the roll test, slowly push the ribbon over the edge of the work surface at the curved radius. Again, measure the length of the piece that breaks off. If the piece is shorter than 120 mm (4¾") the clay content is too low. If the piece is longer than 180 mm (7"), the clay content is too high.

Repeat at least five times. The samples in this type of test are especially susceptible to poor mixing — the break often occurs at a location where there is a pocket of silt or sand or the mix is insufficiently wetted. Because of this, you may be able to justify throwing out the lowest test values and using the three highest ones to get an average value for design purposes.

Raw or Stabilized Jar Test: (particle size by sedimentation) This test gives a rough estimate of the proportions of your soil in terms of gravel, sand, silt, and clay. The resolution is not very fine on this, and it is unwise to rely on jar tests alone, as the proportions of silt and clay are often difficult to determine. That said, it is a relatively simple and economical test that doesn't take very much time. Running multiple jar tests on a single sample can help refine the resolution of the results.

Equipment:

- glass jar with flat bottom and straight sides that has a tightly fitting lid
- a measuring tape
- a timer

- a marker
- potable water
- a pinch of table salt
- a pen and notebook

Procedure:

- Mark the one-third height point on the side of the jar.
- Fill the jar with your soil mix to just above the one-third mark and then compact the soil slightly. Remove any soil above the one-third mark.
- Add water until the jar is two-thirds full. Do not fill the jar.
- Add the pinch of salt and seal the lid of the jar.
- Shake the jar vigorously and let stand for 1 hour.
- At the end of one hour, again shake the jar vigorously, set it down and time 1 minute. After 1 minute, without moving the jar, mark the point at which the soil mix has settled. Note this as T1 — this is the gravel and sand fraction.
- Keep timing, and after 30 minutes make another mark at the point where the soil has settled. Note this as T2 — this is the gravel, sand, and silt fraction.
- After 24 hours, the clay should have all settled out, and T3 should be the same as the one-third height mark on the jar — this is the gravel, sand, silt, and clay combined.

Analysis:

Depth of clay = (T3 − T2)

Depth of silt = (T2 − T1)

Depth of sand and gravel = T1

The clay fraction (%) = [(T3 − T2)/T3] * 100

The silt fraction (%) = [(T2 − T1)/T3] * 100

The sand and gravel fraction (%) = (T1/T3) * 100

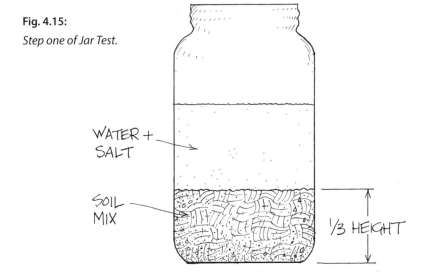

Fig. 4.15: *Step one of Jar Test.*

Fig. 4.16: *Final steps of Jar Test.*

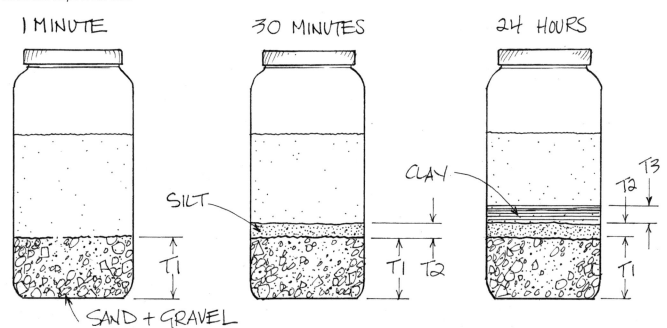

Plotting the proportions on a pie graph can help for comparison with the known acceptable ranges of grain sizes and clay content given earlier.

Raw **Compaction Test:** This test is carried out to determine the optimum moisture content for compaction that will achieve the maximum density with a given soil mix.

To achieve accurate results that are meaningful, consistency in the mix and the tamping action for each sample preparation is essential. This can be accomplished by using the same equipment (e.g. do not substitute a different rammer for tests on the same soil mix) and having the same person carry out the testing throughout.

Equipment:

- scale capable of a minimum 10 kg (22 lb) load, with an accuracy of at least 0.01g
- straight edge or trowel
- square mold 150 mm × 150 mm × 150 mm (6" × 6" × 6"), with a removable extension to contain loose fill above the top before it is compacted. This yields a volume of 0.003375 m^3, or 3.375 liters (0.742 gal)
- rammer with a square-shaped base that is at least 7.5 kg (16.5 lb)
- drying tray(s), oven

Procedure:

- Starting with a very dry mix, which can be labeled "A," fill the mold to between 75 and 100 mm (3" to 4") depth, and then tamp using 18 heavy blows, dropping the rammer from a consistent height each time. Repeat this twice, creating two lift lines within the sample. The third time, remove the extension and trim the sample flush with the top of the 150 mm (6")

Fig. 4.17: *Compaction Test: cubes with increasing water content.*

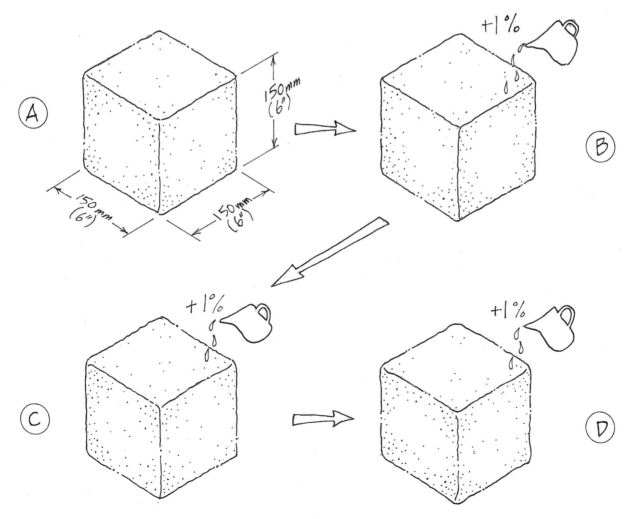

form. Remove from the mold and weigh immediately. Record this weight as AW1. Keep the mix pile covered to keep the moisture content as consistent as possible between steps.

- Taking care not to lose any material, transfer the sample to a drying tray and place in an oven at 80°C (176°F) for 24 hours or until there is no change in weight of the sample. Record the final weight as AW2. The weight of water contained in the original mix is the difference between AW1 and AW2. Dividing the weight of water by the volume of the sample gives the moisture content as a percentage.

- Repeat the test, adding 1% of AW2 mass of water to the original soil mix and blending thoroughly. (This should be about 60 grams, or 2¼ ounce.) Label this sample "B."

- Continue this trend, adding another 1% of water to the mix each time, labeling each sequentially, C, D, etc. Repeat until the dried weight of the sample begins to drop. The sample with the highest dried weight corresponds to the optimum moisture content for compaction.

Note: This is a basic version of the standard *Proctor test*, and when you find the optimum moisture content — and corresponding dry density — you will have found the 100% Proctor density for that particular soil mix. Geotechnical engineers commonly use the Proctor scale to specify density for fill materials that are going to be compacted during construction. A common Proctor test number for any fill that is sensitive to long-term settlement is 95%. Certain situations will require more compaction, and some will be okay with less. For your raw rammed earth walls, the closer you can get to 100% of Proctor, the better they will perform.

Optimum moisture content for stabilized rammed earth is also very important, but there are other factors that come into play. In terms of final compressive strength, the percentage of cementitious binder in the mix makes a larger impact than any increase in compaction above 95% Proctor. That is, an increase from 6% to 8% binder by weight will make a bigger difference than going from 96% to 98% density on the Proctor scale. The addition of more water than is necessary to hydrate the cementitious binder can lead to lower strengths and excess cracking. It also has a strong effect on surface finish — if there is too much water in the mix, it tends to form a paste with the cementitious material, which ends up filling pores and pressing against the formwork to create more of a concrete appearance rather than the sandy, porous surface texture of stabilized rammed earth.

During installation, stabilized rammed earth is much more sensitive to temperature, humidity, exposure to wind and sun, and time. Once the cementitious binder is added, you have effectively started a countdown that does not forgive. Generally speaking, you have one hour to finish mixing, placing, and tamping any given batch. Continuing to add water and mixing will only result in a weaker final product.

You will need to take into account the ambient temperature and humidity at the time of your ramming. Mixes made and tamped in the heat of August will set up much more quickly than those made in the cool of November. You can vary the moisture content to accommodate environmental conditions within the range of acceptable overall performance.

Raw Shrink Box Test: This test is carried out to determine the volumetric stability of the clay portion of the soil in question. It will be useful if you are not certain that the clays in your area are commonly used by potters, natural builders, or plasterers. Note that this test is designed to measure *horizontal* shrinkage, which is critical for crack control. Vertical shrinkage will be lower because of the effect of tamping and the nature of *an-isotropic* materials like soil.

Equipment:

- an open-topped wooden or metal box 40 mm × 40 mm × 600 mm (1⅝" × 1⅝" × 24")
- release agent
- measuring tape
- tamper with 40 mm × 40 mm head
- drying oven

Procedure:

It is important that your tamper be able to fit into the box — whatever its size — so if you have a 52 mm × 52 mm (2" × 2") tamper, then your box should be at least large enough to accommodate the tamper. You can make a wooden tamper for this test at the same time you make the box.

> *Isotropic* means having the same characteristics throughout. Soil is decidedly *an-isotropic*; it can be quite random in terms of material distribution — although at a large-enough scale, it may appear more uniform. When we compact soil by tamping, we are forcing some isotropic character into the mass by physically aligning particles into a more uniform horizontal orientation.

- Starting with a mix at the optimum moisture content determined in the compaction test, fill the box to level with the top, then tamp the soil firmly, using the tamper as consistently as possible, working until it does not compact any more.

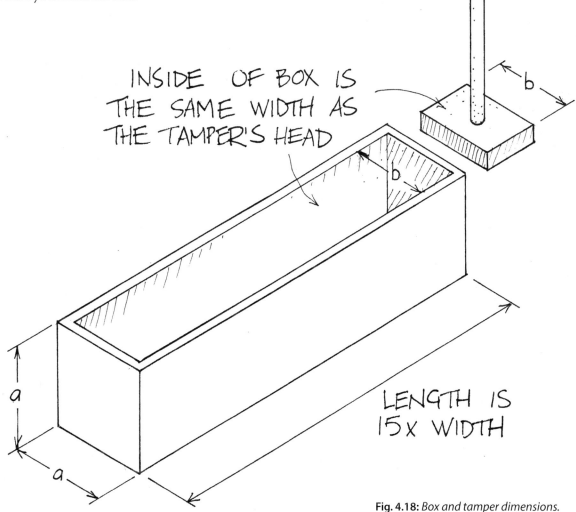

Fig. 4.18: *Box and tamper dimensions.*

- Dry the sample in the box in an oven at 80°C (176°F) for 24 hours or until there is no change in weight of the sample. It is important to dry the sample relatively rapidly for this test to be accurate. Depending on ambient humidity (just allowing it to sit and dry out slowly) may not result in the development of as many cracks.
- After the sample is dry, push all of the soil to one end of the box. You may have to consolidate separated material, as cracks will likely develop in several places as the soil dries out and shrinks. Measure the distance from the end of the box to the dried soil.
- If the shrinkage is 12 mm (½") [2% reduction in length] or less, you may need more clay if you plan to build without stabilizer. Based on this amount of shrinkage, and in consideration of other test results, the soil mix should be fine if stabilizer is being used.
- If the shrinkage is between 12 mm and 24 mm (½" to 1") [2% to 4%], you will need to add stabilizer or a clay-free soil to the mix to avoid excessive cracking.
- If the shrinkage is more than 24 mm (1") [> 4%], you definitely need to add clay-free soil to the mix, whether you are using chemical stabilizer or not.
- If you have used another size of box, use the percentages for your results. The box size given here is a minimum recommended size; larger sizes will work fine, but smaller ones will require great precision to achieve meaningful results.

Raw or Stabilized **Test Walls/ Prisms:** Tests to check on color, texture, and (potentially) the strength of a given mix. See the section on prism geometry in the laboratory tests section below for size of sample that can be used for compressive strength testing, if desired or necessary.

Fig. 4.19: *Consolidating material and measuring shrinkage.*

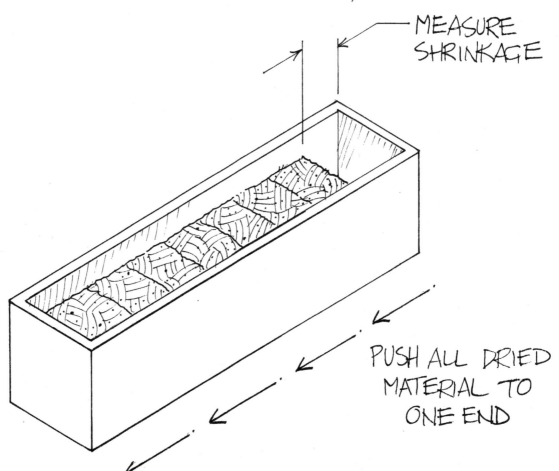

Perhaps no other test is as useful as creating small wall samples — mixing, wetting, and ramming your mixes, even at a small scale, gives you a good feel for the optimum moisture content for ramming. The results of a test wall can tell you how the surface of potential formwork material will impact a given mix, will give you an opportunity to test out different tamper heads and/or techniques, will show you lift lines, and will give you a surface that light plays off of in a similar fashion to the final wall. It's just plain as close to a dress rehearsal as you're going to get. It's a great idea to start off with small garden walls or other non-structural elements before attempting a full-scale wall.

Laboratory tests

There are several versions of the shop tests listed above that can be readily performed by geotechnical laboratories. These include grain-size distribution (both by sieve and hydrometer methods), Atterberg limits, optimum moisture content and compressive strength tests to any one of CSA A23.2, CSA S304.1, ASTM C39, ASTM C873, or AASHTO T22 standards. It is quite likely that your building official will require testing carried out by a third-party laboratory to a specified, code-referenced standard.

For the grain-size distribution, Atterberg limits, and optimum moisture content tests, it is sufficient to provide an adequate volume of the soil mix or mixes you want tested — clearly labeled to avoid confusion. However, for compression strength testing, it is up to you to prepare the samples beforehand, in accordance with the specification for the standard being tested to. There are two sample geometries relevant to compression testing: cylinders and rectangular prisms.

Cylinders

Cylinders are by far the more common style, and all the relevant standards have a common height-to-diameter ratio — 2:1. The most common concrete test cylinders tend to be 150 mm (6") diameter and 300 mm (12") in height. This

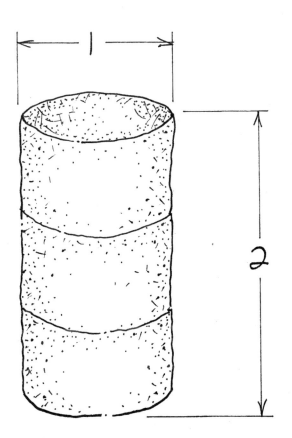

Fig. 4.20: *Sketch of rammed earth test cylinder proportions.*

4"(100mm) ⌀ x 8"(200mm) HIGH CYLINDERS ~ 6 TO 8 lbs.

6"(150mm) ⌀ x 12"(300mm) HIGH CYLINDERS ~ 20 TO 25 lbs.

is a suitable size for construction phase testing, but if you are conducting extensive tests on a variety of soil and stabilizer mixes, it adds up to quite a bit of material in a short period of time. The minimum cylinder test sample size, and one that is quite common for AASHTO and similar road construction testing protocols, is 100 mm (4″) diameter by 200 mm (8″) high.

In either case, if you are using cylindrical forms designed for conventional concrete construction, you will need to add a section to the top of the form in order to allow for the bulk, low-density, uncompacted earth mix to be added for the top layer and then tamped down to the final height of the cylinder. Some rammed earth builders simply make extra long forms that allow some extra overall height which can then be cut down. For high-strength mixes that require rigorous testing before construction — say, to satisfy a very demanding building department or for a design with some ambitious geometry — it may be beneficial to make the cylinders at least 25 mm or 50 mm (1″–2″) taller than the final height. Then, after curing, the lab technicians can cut the top with a saw and get a very smooth surface for fitting into their test rig. Cutting does not always yield a smooth, level surface, and in these cases, the labs can employ grout to ensure proper seating of the sample at both the top and bottom. This helps eliminate "noise" in the test data caused by seating the cap of the compression test equipment, as well as small cracks and spalling which will reduce the ultimate compressive strength numbers. Some testing laboratories employ a fast-setting sulfur plaster cap on the bearing surface of cylindrical samples that aren't perfectly smooth and level to get a consistent seating across a full set of tests.

It is also possible to make the cylinders a bit short and apply the fast-setting cap material to either the top or bottom (or both). This may be necessary with lower-strength, crumbly samples. Find out what your test lab's preferred methods are and follow those. Individual cylinder crush tests can run at $50 each (or more), and if you have to transport your samples any distance, it can become a large time and money investment quite quickly.

Table 4.4 is a summary of some key characteristics of two sets of theoretical test cylinders. These rough calculations show that the necessary test machine capacity for a 15 MPa mix in 6″ diameter cylinders is around 274 kN, or 62,000 lbf.

Sample storage — while curing, the samples should be covered to prevent rapid loss of moisture. Ideally, a "damp cure" is carried out, which entails wrapping the sample with wet cloth for the first week after it is made. This simulates the initial cure time when the formwork is still in place and the majority of the surface area of the wall system is not exposed to air.

It is a good idea to store samples on a pallet for ease of transfer between site and test lab — assuming you have access to equipment that can easily lift and carry large loads in this manner.

Prisms

Prisms are rectangular samples conforming to masonry design standards. In jurisdictions requiring strict conformance to published

Table 4.4: Summary of cylindrical test specimen characteristics

Cylinder diameter, mm (inches)	Area, mm² (in²)	Compressive strength, MPa (psi)	Weight, kg (lb)	Force required to crush, kN (kip)
100 (4)	8,100 (12.6)	7.5 (1,100)	2.7 to 3.6 (6 to 8)	61 (13.7)
100 (4)	8,100 (12.6)	15 (2,200)	2.7 to 3.6 (6 to 8)	122 (27.4)
150 (6)	18,200 (28.3)	7.5 (1,100)	9.1 to 11.4 (20 to 25)	137 (31)
150 (6)	18,200 (28.3)	15 (2,200)	9.1 to 11.4 (20 to 25)	274 (62)

design standards, it may be necessary to create these (rather than cylinders) for compressive strength tests. On the one hand, these types of samples are more representative of the actual built geometry, and the test results are less influenced by edge effects. However, to achieve the minimum prism size required by the CSA S304.1 test protocols in Annex D, a 150 mm (6″) wide sample would need to be at least 300 mm (12″) long and 300 mm (12″) tall. This is based on the assumption that a 150 mm (6″) wide wythe is sufficient for the wall design in question. The minimum width of the prism is the minimum width of the wall required for the load and geometry being proposed. The recommended height-to-thickness ratio for prisms in the CSA standard is 5. That would mean a 750 mm (30″) tall prism — which is likely to be unpractical. A compromise allowed by the authority having jurisdiction in one case was to go with a ratio of 2.5 height to thickness, which is what our example shows in Figure 4.21. The minimum height-to-thickness ratio given in the CSA standard is 2, but reducing the ratio from the recommended 5 means a reduction in the calculated design strength. Going down to a sample of this height means a 15% reduction in design strength from the results of the testing. This is because a shorter, squatter section is less slender, and buckling behavior is a critical failure mode for compression elements. So, a sample made of the same material or mix that is shorter may appear to have higher ultimate strength, and may give a false impression of the actual characteristics of the built wall.

Even if the penalties for using a shorter prism are acceptable in the design calculations, the smallest acceptable prism will be very heavy and awkward to move around. It is also not a simple thing to find a laboratory with a compressive test apparatus tall enough to accommodate samples higher than 400 mm (16″). That, and a 150 mm × 300 mm surface area (6″ × 12″), means that the capacity of the test rig needs to be quite large.

Again, some rough calculations, summarized in Table 4.5, show that not only do we need a test machine capable of accepting a sample

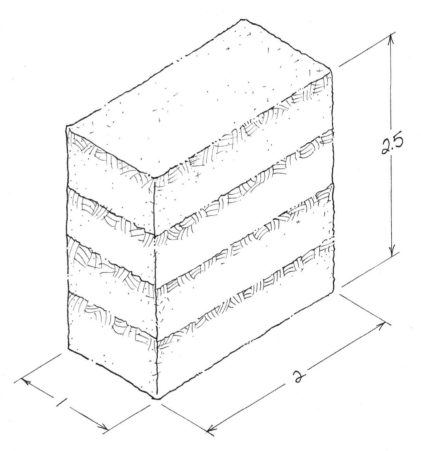

Fig. 4.21: Sketch of masonry prism showing ratios of width, length, and height.

Table 4.5: Summary of prismatic test specimen characteristics

Prism dimensions, mm (in)	Area, mm² (in²)	Compressive strength, MPa (psi)	Weight, kg (lb)	Force required to crush, kN (kip)
150 × 300 × 375 (6 × 12 × 15)	46,450 (72)	7.5 (1,100)	31.8 to 36.3 (70 to 80)	349 (79)
150 × 300 × 375 (6 × 12 × 15)	46,450 (72)	15 (2,200)	31.8 to 36.3 (70 to 80)	698 (158)

over 375 mm (15″) tall, even a 7.5 MPa mix requires almost 350 kN (nearly 80,000 lbf) minimum compressive load capacity. I can say from personal experience that these types of test machines do exist, but they are not common. In one case, a client had to drive over 400 km (250 miles) with samples to get to the nearest lab both willing and capable of crushing the prisms.

Construction Phase Testing

You will likely want to come back to this section after going through the rest of the chapters. These tests apply to raw and stabilized rammed earth alike.

Ball Drop Test: Checks moisture content for optimum compaction during mixing. Ideally, the mixing process is well-enough controlled to keep the moisture added to the mix close to the optimum content for maximum compaction. However, it is good practice for the crew responsible for the mix to check each batch with a quick ball test before delivering the mix to the forms.

As mentioned earlier, for stabilized rammed earth, the "ball" part of the test is more important than the "drop" part of the test. This is because lower moisture contents are more likely to be appropriate for stabilized rammed

Fig. 4.22a and b: *Two views of an SRE prism in machine after compression testing at a lab.* PHOTO CREDITS: TERRELL WONG

earth than for raw, and the ball will shatter when dropped. Forming the ball with stabilized mix means making sure that there are clumps being created when the material is pressed together. Anecdotally, there are stabilized rammed earth mixes that have achieved over 20 MPa that failed the drop test — because finishing requirements dictated using a mix with a lower-than-optimum moisture content. This means that a dropped ball of stabilized rammed earth may shatter in a way that for a raw rammed earth mix would be a disaster. Remember that the mix will feel different after binder has been added.

In Ontario, materials from aggregate suppliers are coming in at 3% to 4% moisture content. It's good to do tests to get a feel for what 4, 6, 8, and 10 percent moisture content feels like. This can be done during initial strength and texture testing, but also during construction. Ideally, you should learn the *feel* of different moisture contents, because you will not have time to run exploratory tests while ramming. Each batch will have to "feel" the right way. During the design test phase, it is useful to do a stabilized version of the compaction test along with test walls and cylinders in order to coordinate the ideal moisture content for strength, texture, and finish. At the very least, dry out a volume of your mix overnight in an oven, then add water, incrementally increasing the amount and literally feeling the material each time. Do some ball/drop tests so that you are comfortable with what the mix should feel like when you are working at full scale. Again, when you have added binder, the mix will feel different in your hand than raw soil does — a good stabilized mix is usually quite a bit drier than a raw mix.

Cover materials to try to preserve moisture as much as possible. Water needs to be added for binders, pigments, and for silicate emulsions or other internal sealers, but keeping the amount added to a minimum helps with mix consistency. The water that is already in the mix should be relatively evenly distributed, and this is helpful for mixing. In a 1,000 kg (2,204 lb) batch of stabilized rammed earth, you may only be adding 20 to 30 kg (44 to 66 lb) of water.

Compressive Strength Tests: Similar to design phase lab tests, but samples are created during ramming and tested to confirm that the as-built wall has the minimum strength required by the design. Cylinders are the simplest samples to use for construction phase compressive strength tests.

It is good practice to make test cylinders for each mix — if not each batch. A bare minimum would be making test samples on every day that ramming is taking place. Whenever test cylinders are made, at least two should be done, carefully noting the recipe, date, and any other detail that will help to identify the location in the building represented by each particular sample. Making at least two samples at a time allows for testing of the given batch at both 28- and 56-day cure times. More may be necessary if 7- or 14-day strengths are required. (This can be the cases when a construction schedule is compressed and high point loads from large steel beams or large lateral loads from something like backfill need to be applied very shortly after the wall is rammed.) Having multiple test results with the only variable being cure time gives an idea of the rammed earth's long-term compressive strength, how fast the strength is developing, and how long it is taking to achieve this strength.

In some cases, you may be required to provide a bored sample out of a built wall section. Concrete coring tools are available for this task. If it is not possible to take the sample from the top of a wall section, it can be taken out of the side of the wall. The location of the borehole can be coordinated with utility access points or simply taken from somewhere that is not critical. An advantage of sampling an in-situ wall is that

Fig. 4.23: Test cylinder split molds made from 100 mm (4") dia. PVC pipe and hose clamps. Note extra length to allow proper tamping for upper lift.

Fig. 4.24: Manufactured test cylinder split molds, Australia.
Photo credit: Christopher Beckett

Barker single family residence, in the Muskoka region of Ontario, Canada. Stabilized rammed earth walls are by Tapial Homes.
PHOTOS ARE BY EMILY BLACKMAN.

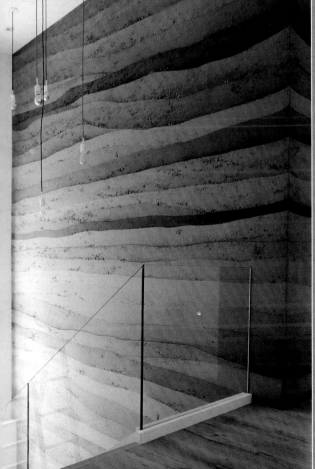

Allen single family residence in Huntsville, Ontario, Canada. Stabilized rammed earth walls are by Muskoka Sustainable Builders.
PHOTO BY DALILA SECKAR.

Single family residence in Chelsea, Quebec. Stabilized rammed earth walls are by Bautechnik.
PHOTOS BY: BAUTECHNIK (REINECKE).

Top: *Haliburton Solar and Wind showroom, in Haliburton, Ontario, Canada. Stabilized rammed earth walls are by the Sandford Fleming Sustainable Construction Class of 2016, under the supervision of Tapial Homes crew.* PHOTO IS BY EMILY BLACKMAN.

Bottom left and right: *Smyth single family residence in Kemptville, Ontario, Canada. Stabilized rammed earth walls are by Aerecura Builders.* PHOTOS ARE BY DALILA SECKAR.

Hanson single family residence in Castleton, Ontario, Canada. Stabilized rammed earth walls are by Aerecura Builders.
PHOTOS ARE BY TIM KRAHN.

Above and Left: *Huck bunkie on Browning Island, Lake Muskoka, Ontario, Canada. Stabilized rammed earth walls are by Muskoka Sustainable Builders.* Photos are by Dalila Seckar.

Below left and right: *Telenor Tower near Islamabad, Pakistan. The biggest rammed earth project in the world. By SIREWALL.* Photos are by Rana Atif Rehman

McCully single family residence in Huntsville, Ontario, Canada. Stabilized rammed earth walls are by Muskoka Sustainable Builders.
PHOTOS ARE BY DALILA SECKAR.

Top: *Oxford County Waste Management & Education Centre in Salford, Ontario, Canada. The centre was opened in 2018 and is the first municipally owned and operated net-zero building in Ontario. Stabilized rammed earth walls are by Tapial Homes.* PHOTOS ARE BY TIM KRAHN.

Center left and bottom: *Southeast Wyoming Welcome Center. Rammed earth construction in subfreezing temperatures. Commercial scale, cured, load-bearing walls and deep backfilling.* PHOTOS ARE BY AZZARELLO PHOTOGRAPHY

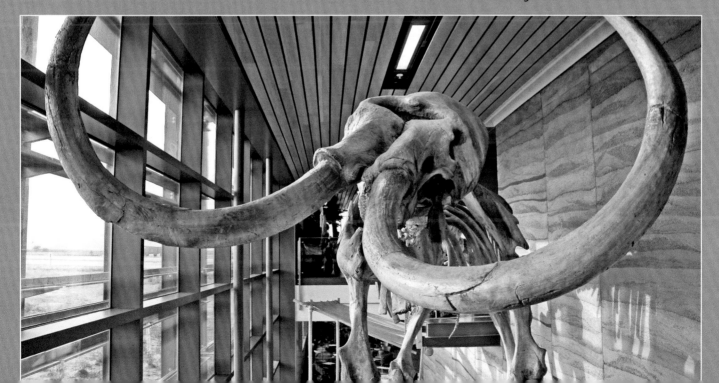

A single family residence on Salt Spring Island, British Columbia, Canada. Stabilized rammed earth walls are by Clifton Schooley and Associates.

PHOTOS ARE BY CLIFTON SCHOOLEY.

the tested material should accurately reflect the actual constructed wall.

Bored samples can be taken either vertically or horizontally. Vertical samples can be taken at the top of wall sections, and then the void can be refilled before finishing the top of the wall. The disadvantage is that you don't always have access to the top of a wall after the building is closed in. Horizontally bored samples can be taken after the roof is on. A disadvantage of taking the sample horizontally is that the vertical axis of the test cylinder is perpendicular to the direction of ramming, and these cylinders will give a different interpretation of the compressive strength of the material. That said, it can be valuable to have both — and horizontally bored cylinders can be tested for tensile strength when crushed across the cylindrical axis. To date, this type of testing has not been required in any jurisdiction where I have worked — primarily because design considerations neglect any tensile capacity in the rammed earth.

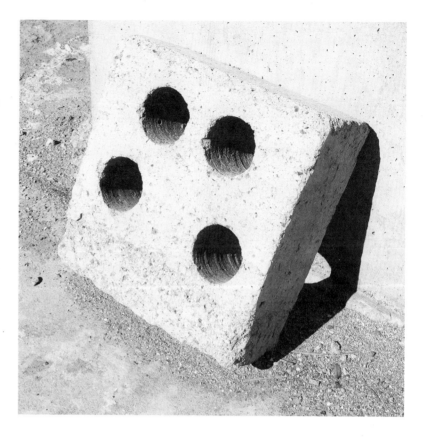

Fig. 4.25: *Sample block with bored cylinders removed.*

Photo credit: Dalila Seckar

Relevant Research

Bhattacharja, Sankar, et al. "Stabilization of clay soils by Portland cement or lime: A critical review of literature." PCA Rand D Serial No. 2066, Portland Cement Association, Skokie, Illinois, 2003.

Cheah, J.S.J., et al. "Evaluating shear test methods for stabilised rammed earth." *Proceedings of Institution of Civil Engineers, Construction Materials* 165(6) 2012, pp. 325–334.

Hall, M. and Y. Djerbib. "Rammed earth sample production: Context, recommendations and consistency." *Construction and Building Materials* 18(4) 2004, pp. 281–286.

Jaquin, P.A., et al. "The strength of unstabilized rammed earth materials." *Geotechnique* 59(5) 2009, pp. 487–490.

Khalife, Roy, Pranshoo Solanki, and Musharraf M. Zaman. "Evaluation of durability of stabilized clay specimens using different laboratory procedures." *Journal of Testing and Evaluation* 40(3) 2012, pp. 363–375, ASTM.

Smith, J.C. and C.E. Augarde. "A new classification for soil mixtures with application to earthen construction." ECS Technical Report, Durham University, 2013.

Windstorm, Bly and Arno Schmidt. "A report of contemporary rammed earth construction and research in North America." *Sustainability* Vol. 5, 2013, pp. 400–416.

Chapter 5
Wall System Examples and Structural Design Considerations

There are many possible types of rammed earth wall assembly: monolithic, insulated or uninsulated, multi-wythed, thermally isolated, and more. The design that I have been working on for the past decade or so is the internally insulated, double-wythe profile pioneered by Meror Krayenhoff of Salt Spring Island in British Columbia. Variations on this design are shown later in this chapter, but first, we'll run through some other construction styles.

A monolithic wall is a single unit, and it can be either raw or stabilized, with or without mechanical reinforcement. A basic monolithic wall section is shown in Figure 5.1. As will be laid out in the design considerations given later in this chapter, a raw rammed earth wall needs to be nearly 300 mm (12″) thick to be considered structural; for code compliance, it will need protection from weather if it is on the exterior of the building. That said, in order to take advantage of rammed earth's thermal mass, externally applied insulation (*out-sulation* to some practitioners) makes a lot of sense. So, an externally installed insulation layer and cladding over that can make a good wall assembly. Figure 5.2 is a section sketch of an externally insulated monolithic rammed earth wall.

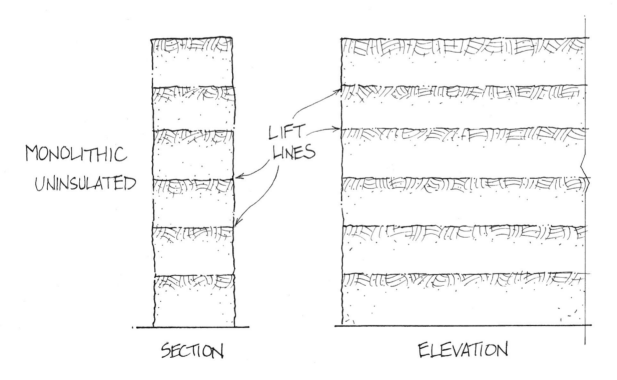

Fig. 5.1: *Monolithic, uninsulated rammed earth wall in section and elevation views.*

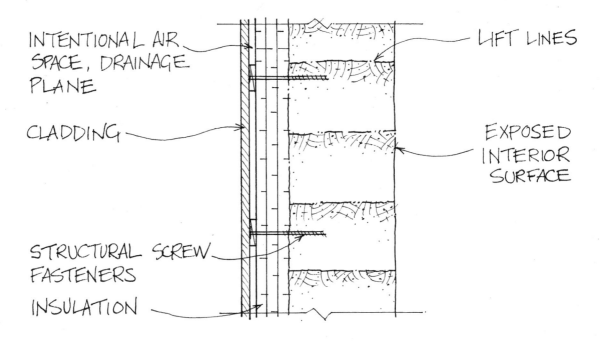

Fig. 5.2: Externally insulated monolithic rammed earth wall section.

What about the base of the wall? A typical foundation for rammed earth walls in North America is the concrete strip footing. These footings may or may not have a shear key, as shown in Figure 5.3, and will likely have internal reinforcement as well as a direct connection into the rammed earth itself, as shown in Figures 5.3 and 5.4.

Because of the dynamic pressures exerted while tamping a rammed earth wall, and the extensive formwork structure necessary to keep the material contained, their foundations need to be stiff and strong. The reinforced concrete typically used has a high thermal conductivity, similar to rammed earth itself. This presents a challenge for energy-efficient designs that also aim to minimize embodied carbon. I have worked with builders to create thermally broken footings, and on designs with externally insulated "floating" slabs rather than strip footings.

In cases where piles are necessary, it may be possible to be especially innovative. On one such project, a concrete grade beam cast in place above concrete piles was thermally broken at the top to match the internally insulated rammed earth double-wythe wall above. Figure 5.4 is a detail from the plan set on that project. A major feature of this foundation, not shown in the section provided, is the incorporation of hydronic loops in the cast-in-place piles. The site, in Winnipeg, Manitoba, had highly plastic clays extending more than 15 m (50′) below grade, above glacial till and fractured limestone bedrock. The 450 mm (18″) diameter piles were spaced on 2.5 m to 3 m (8′–9′) centers, and there were a total of 20 of them around the perimeter of the building. The mechanical engineer and I worked out a construction sequence to install hydronic lines that allowed the use of the concrete in these piles as the heat source/

sink field for a ground source heat pump.

The wall section detailed in Figure 5.4 specifies the use of foil-faced rigid insulation on the interior side and for the joints between sections of insulation to be taped and sealed. This is to conform explicitly with building code regulations requiring a proven vapor barrier. I can tell you that, practically speaking, it is nearly impossible to tape sections of insulation together within a set of formwork while tamping a rammed earth wall. I can also tell you that 150 mm (6″) of rammed earth is a more effective vapor retarder than 6 mil polyethylene, for reasons laid out in Chapter 3.

An example of an *alternative solutions* proposal for a nonconforming wall assembly due to untested vapor permeance is given in Appendix B.

There are many possibilities for advancing the design and construction of foundations. Low-carbon options for concrete-free foundations (or low-carbon concrete mixes) are becoming available, and I encourage the reader to do some research before "settling" for a conventional reinforced concrete strip footing. Properly constructed rubble trench foundations can support rammed earth walls. Stabilized rammed earth works below grade and can function as basement walls. Earthbag footings hold a lot of promise, as do helical piles and wooden pier foundations.

What about the top of the wall? It's not easy to create a smooth, level surface of rammed

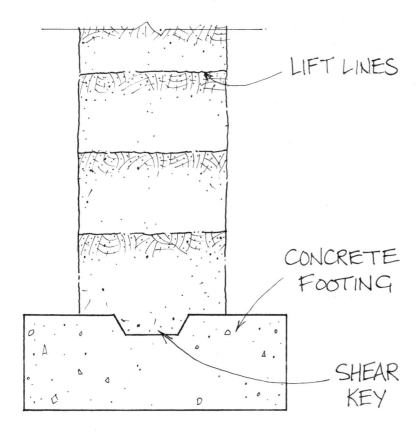

Fig. 5.3: *Monolithic rammed earth wall to concrete footing connection in section.*

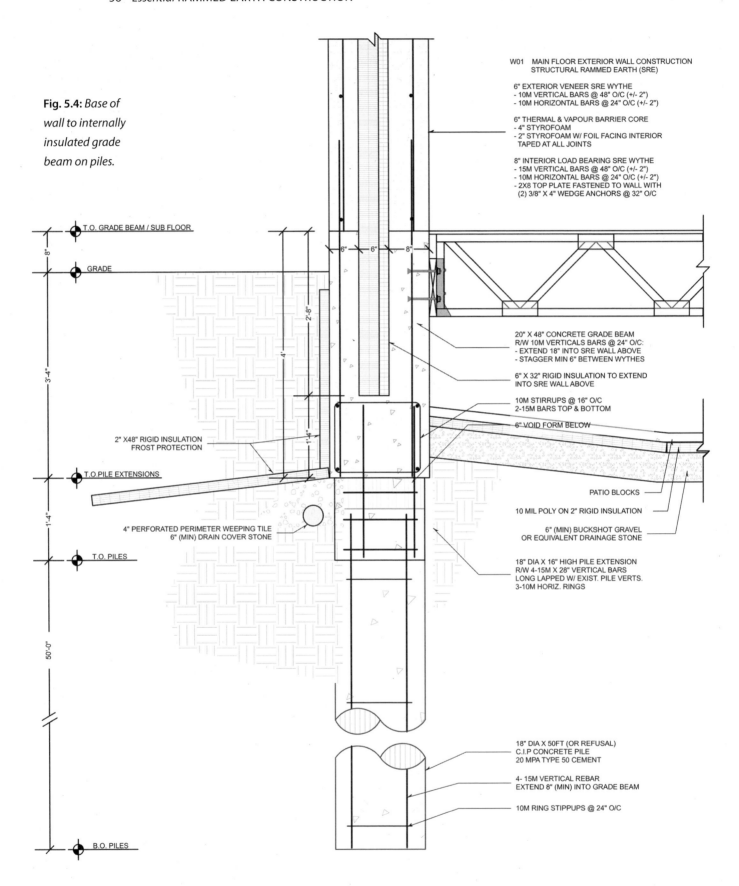

Fig. 5.4: Base of wall to internally insulated grade beam on piles.

Fig. 5.5: *Formwork for cast concrete bond beam.*
Photo credit: Kevin Smyth

Fig. 5.6: *Cast bond beam with screed finish and threaded rod for anchoring wood sill.* Photo credit: Kevin Smyth

earth using pneumatic tampers — or even hand tampers, for that matter. Instead, it is common to create some kind of bond beam at the top of rammed earth walls, either of a wet mix from the same material as the wall, placed and troweled smooth by hand, or using concrete, as shown in Figures 5.5 and 5.6, or with timbers as shown in Chapter 9, on finishes.

The concrete bond beam shown in Figures 5.5 and 5.6 was actually placed as a repair — the stabilized rammed earth at the top of the wall wasn't mixed or placed properly, and it spalled away badly not long after initial construction. The owner drilled and grouted threaded rod down into the competent rammed earth lower down, and used plywood to create a temporary formwork after the rest of the wall was completed. The plywood was set to an appropriate elevation to screed level, and the concrete was placed, as shown in Figure 5.6.

It may be that you do not want to put a roof or floor system above your rammed earth wall. Perhaps you are designing a feature wall for your garden, as shown in Figure 5.7. You may not want the wall to be straight or level at all — and

Fig. 5.7: *Curved garden wall. Note the way that precipitation is running down the face of the wall, following cold joint lines for portions of their path to the ground.*

Fig. 5.8: *Close-up of textured, sloped top on curved garden wall. This design is intended to direct precipitation down slope while minimizing the amount of water that flows down the face of the wall.*

it doesn't need to be. Figure 5.8 shows the sculpted top layer of the garden wall, designed to minimize precipitation running down the face of the wall; the textured surface directs rain and snow melt down the top of the structure.

Curved walls are a challenge to form, to say the least. I won't go into any detail on curved walls other than to say that they are possible, but they should be planned carefully and ideally worked out at scale or on practice structures first. Figure 5.9 is a photo of a curved practice wall that David Easton's crew did at their shop before executing the final design for a client.

Figure 5.10 is a symmetrical double-wythe wall with 150 mm (6″) stabilized rammed earth wythes on either side of a 150 mm (6″) rigid insulation core. The two rammed earth wythes are

Fig. 5.9: *Curved test wall in Napa, California. Note the serpentine top template leaning against the shop wall.* Photo credit: Dalila Seckar

tied together using high-density polyethylene (HDPE) geogrid, spaced to coordinate with the insulation sections at 600 mm (24″). HDPE was chosen to minimize thermal bridging across the internal insulation. It functions as designed, but was not the easiest material for the builder to work with. Other low-conductivity materials used for inter-wythe connection include glass fiber reinforced polymer (GFRP) rod, fiberglass grids, and basalt-based rebar. These products tend to be expensive and difficult to source. It is not always easy to find structural characteristics for pull-out capacity or development length when working with novel materials, but it may be worth it for the long-term durability factor or the energy-efficiency gains.

Fig. 5.10: *Example wall section with engineering detail; symmetrical, thermally isolated.*

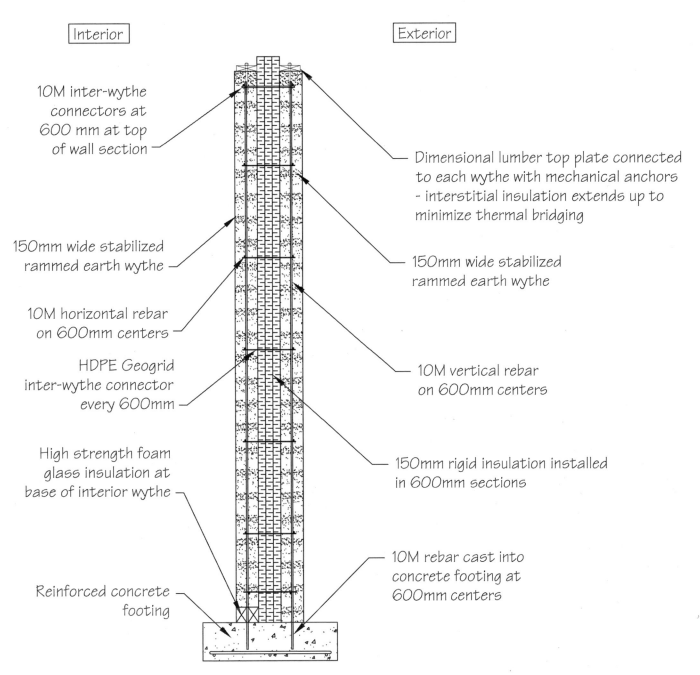

> Mechanical reinforcement is typically some kind of bar, wire, or cable placed within a mass, like a wall or footing, to enhance the strength of the whole assembly. In our case, the rammed earth is the mass and it has good compressive strength, but not so much in terms of tension. Adding rebar, which is very strong in tension, raises the load capacity of our wall.
>
> The ability of the reinforcement to take tensile load depends on the connection between the rebar and the rammed earth. There needs to be sufficient contact area between the two materials, which is defined as the *development length*, or *embedment*.

This wall section also features a structural, thermally isolated internal wythe. The thermal isolation at the base of the wall is provided by foam glass, which is much stiffer than rigid polystyrene. Autoclaved aerated concrete is another product that could be used in this detail. The thermal isolation of the exterior vertical side is via rigid insulation that is connected to the attic/roof insulation at the top of the wall. Thus, the only uninsulated side of the wall is the face on the interior of the conditioned space.

The wall is connected to the concrete footing below with 10M (#3) vertical steel rebar that was cast in place. This isn't always practical, and many builders prefer to install the rebar after the footing has set, during the formwork erection process. Another option is to cast shorter sections during the foundation work and then to tie or weld longer bar to the ends of the embedded bar. Horizontal rebar is placed within each wythe at the same elevations as the inter-wythe connectors.

This wall section has the insulation running up past the top, where it connected to the roof insulation to provide a continuous thermal envelope.

Figure 5.11 is a nonsymmetrical stabilized rammed earth wall. The interior wythe is the sole structural element in the design, and it is 200 mm (8") thick. The exterior wythe is 150 mm (6") thick, and is only considered to be a veneer in this design. The wythes are interconnected with welded wire mesh cut into narrow strips, their location coordinated with the 600 mm (24") wide insulation sections.

Vertical rebar in this example is 15M (#5) for the structural wythe and 10M (#3) for the veneer wythe. The top of the wall has an engineered wood sill plate made of laminated veneer lumber (LVL) that spans both wythes. At 500 mm (18") wide, this does create a thermal bridge — but one that the client was willing to live with, as wood is not as poor an insulator as concrete.

The structural design for this wall required a strong connection between the top plate and the roof system, so there are two horizontal runs of 10M (#3) bar at the top of each rammed earth wythe, and the anchor bolts holding the sill plate down were tied into these bars.

Wall System Examples and Structural Design Considerations 61

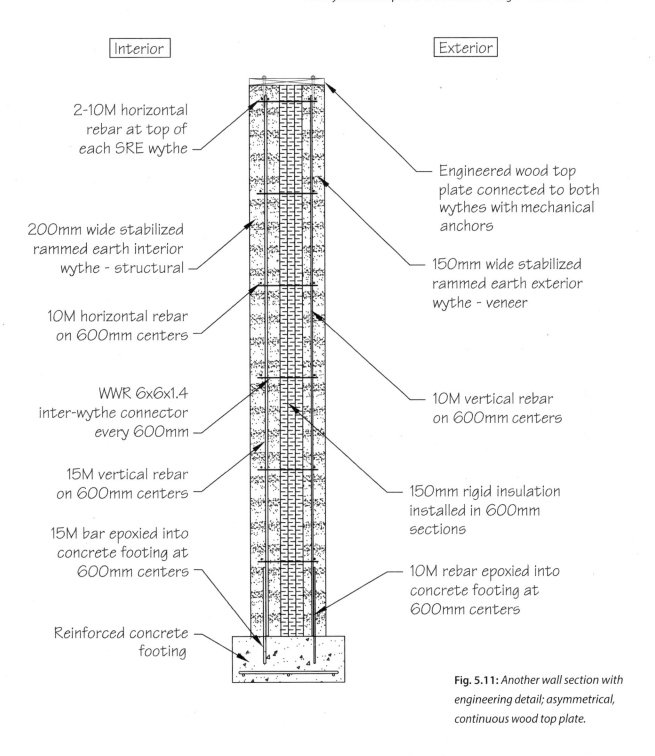

Fig. 5.11: Another wall section with engineering detail; asymmetrical, continuous wood top plate.

Structural Design Considerations

Detailed engineering analysis and design is generally complex and often complicated, some would argue unnecessarily so. With that in mind, several material standards and building codes provide empirical guidelines that simplify the key issues into "rules of thumb" so that buildings within certain size limits and not exposed to extraordinary loading can be safely and economically designed by non-engineers. It bears noting that these sets of empirical structural guidelines are given with the caveat that seismic loading is not covered, and the relatively low building heights mean that wind loading is not likely to be of much concern. In other words, the building structure is only going to be affected by gravity loads, and, due to its size, shape, and location, no large lateral loading will occur. If you are in a location where earthquakes occur, or if your site is subject to extremely high wind loads, you are advised to consult an engineer.

Raw Rammed Earth Design Guidelines

It is not within the scope of this book to get into the finer points of various structural design theories and their practical implications, but the interested reader is directed to Jacques Heyman's *The Stone Skeleton: Structural Engineering of Masonry Architecture*. Heyman's closing statement is of particular relevance: "The key to the understanding of masonry is to be found in a correct understanding of geometry." For our purposes in this book, we are going to rely on empirically determined rules of design based on the correct basic wall geometry for our material.

Units

A word about units, density, force, and specific weight. The unit of mass in SI is the gram (g). In Imperial, it is the pound (lb). The unit of force in SI is the Newton (N). In Imperial, it is pound-force (lb-f). Many people steeped in the use of Imperial units leave off the '-force' suffix and refer to both mass and weight as being in pounds. I find this unnecessarily confusing. Mass is a physical characteristic of all matter, and it does not change. Weight is the force that is the result of an object's mass under acceleration, in our case that of earth's gravity. The earth's gravitational acceleration constant is 9.807 m/s^2 in SI and 32.174 ft/s^2 in Imperial units. A Newton then, is a kg-m/s^2.

For example, a bowling ball has the same mass on earth as it does on the moon, but its weight changes between the two locations because the acceleration caused by the moon's gravity is less than that of the earth's. However, because of the basic definitions used for the Imperial system, a pound of mass is also a pound of force (under the earth's gravitational acceleration). I leave the interested (confused?) reader to research this — the history of weights and measures in both the UK and the US is fascinating, to say the least. There is at least one revolutionary war involved, and a bit of history will provide an explanation of the motivation to come up with a more universal metric.

Density is given as a mass per unit volume (e.g. kg/m^3 or lb/ft^3), while specific weight is the weight per unit volume (e.g. kN/m^3 or lb-f/ft^3).

Perhaps not wanting to be left out of the confusion-causing semantics, the SI system has a special term for pressure, the Pascal (Pa). It is a unit of force exerted over a given area (e.g. N/mm^2). The Imperial system simply refers to pressure in "psi" or "ksi," (lb-f/in^2 or kilo-lb-f/in^2, respectively). And in recent years, practitioners using Imperial units have adopted the SI practice of adding prefixes to abbreviate quantities, e.g. 1,000 lb-f is often referred to as a "kip" or kilo-pound.

I will try to be consistent in my use of units — if your eyes are already glazing over, I apologize. I'm reminded of a geology professor I once knew who said that he and his colleagues studied what they could see, made guesses about what they couldn't see, and then came up with their own vocabulary to make the rest of us think they were more informed than we were.

A reasonable empirical design guide, outlining a set of basic geometries for unreinforced, raw rammed earth can be taken from a summary of Annex "F" from the CSA S304.1 Design of Masonry Structures standard, along with some input from the New Zealand NZ 4297 Earth Building standard; ASTM E2392: Standard Guide for Design of Earthen Wall Building Systems; and MSJC: Building Code Requirements for Masonry Structures. I have tried to lay this out in a format that provides minimum and maximum geometric "rules" for wall design: height, thickness, distance between perpendicular support, opening sizes and locations, and anchorage details.

Taking the case where raw rammed earth is equivalent to rubble stone masonry laid dry, the minimum wall thickness is 290 mm (11⅜"). This is not unreasonable — in many places, the minimum wall thickness for raw rammed earth is set at 300 mm (12"). This is for both structural and thermal requirements. It is important to note that this is a minimum thickness for load-bearing walls. Raw rammed earth walls that are non-load-bearing can be safely built at 200 mm (8") thickness — and even thinner if they incorporate mechanical reinforcement, such as deformed metal bars (rebar), or geometry that increases stability, such as curves or buttresses. However, this sort of design is again moving us out of the empirical world. Not to say it cannot be done — but you are advised to consult with experienced builders and probably a structural engineer who understands earthen materials.

Now that we have a minimum wall thickness, we can define a maximum wall height. This is governed by the slenderness ratio: the height of the wall to its thickness, or H/T. The maximum slenderness ratio from Annex "F" in S304.1 is 14. Therefore, the maximum load-bearing wall

Fig. 5.12: *Sketch of wall section with slenderness ratio H/T = 14.*

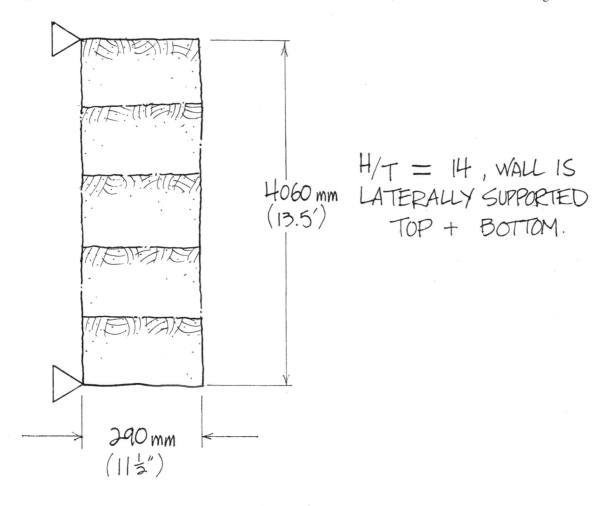

height for this width of wall is 4,060 mm (13′-3¾″). It is important to note that this wall must be laterally supported at both the top and the bottom. In a house, this is usually some kind of floor diaphragm at the base and a roof/ceiling diaphragm at the top.

Loads and Structural Capacity

So, now we have our basic wall section — what can it support? The minimum compressive strength given in the S304.1 standard for this masonry equivalent is 0.7 MPa (~100 psi). This is a realistic, albeit conservative value for well-mixed and compacted raw rammed earth. Many standards point to 1.0 MPa (~145 psi) as a target strength for a well-compacted rammed earth mix. For design purposes, using 0.7 MPa should keep loads well within the capacity of the material, allowing for variations in quality. For basic single-story designs, this will be adequate to handle imposed loads.

For example, a typical raw rammed earth wall with a compressive strength in the 0.7 MPa range will have a density of at least 1,800 kg/m^3 (~112 lb-f/ft^3). This is then the minimum value we should use when calculating the self-weight of the wall. Per meter length of wall, using the dimensions above, we have a volume of 0.29 m × 4.06 m × 1.0 m = 1.18 m^3 (41.7 ft^3). The mass of each meter of rammed earth is 2,124 kg (4,683 lb). Calculating force from mass, 2,124 kg × 9.807 m/s^2 = 20,830.1 N. This is the rammed earth in our wall's own dead load. Most engineers are fond of working with numbers in tripled orders of magnitude, e.g. 1,000 g = 1 kg or 1,000 lb-f = 1 kip. Engineers are also fond of rounding and keeping things on the conservative side, depending on the level of precision and accuracy warranted by their knowledge of the materials and assembly in question. So, I would say that we have a self-imposed dead load of 20.8 kN (4,683 lb-f). Some colleagues would probably say that there too many significant figures in that number, that I should just use 20 kN and move forward knowing that I'm being conservative.

The area this load is applied over is 0.29 m^2 (3.12 ft^2), and at 0.7 MPa (100 psi) compressive strength, this footprint has an unfactored load capacity of 0.29 m^2 × 700,000 N/m^2 = 203,000 N, or 203 kN (45.6 kip). So, we have a "spare" capacity of around 203 − 20.8 = 182.2. So we could say we have 180 kN/m (13.2 kip/ft) to work with when considering other loads on the wall.

If our wall is supporting a light wood truss roof with a span of 9.15 m (30′), and we are in a climatic zone with a ground snow load of 1.6 kPa (33.4 psf), a wet snow load of 0.4 kPa (8.4 psf) and a minimum dead load of 0.5 kPa (10.0 psf) — then our unfactored line load on top of the wall is:

Snow:
([1.6 kN/m^2 + 0.4 kN/m^2] × 9.15 m) / 2 walls = 9.15 kN/m (675 lb-f/ft) along the length of each bearing wall

AND

Dead:
(0.5 kN/m^2 × 9.15 m) / 2 walls =
2.29 kN/m (169 lb-f/ft) of dead load.

So, our remaining capacity is still just under 170 kN/m (11.5 kip/ft), or above 90% of the strength available. We should not get too excited at this result — no load factors have been applied. Load factors are applied to both sides of these equations — raising or lowering the resistance of a given structural unit and the force applied to account for a specific situation. The numerical value of the load factors are based on geometry (e.g. slope of a roof), material qualities (e.g. if the roof is slippery), location (e.g. coastal Maine vs Austin, Texas), duration of load, and a number of other statistically relevant inputs.

I know I'm risking confusion by bringing partial engineering calculations into this, but it is my hope that by doing so I will show readers the importance of staying within the empirical guidelines — and the equal importance of consulting an engineer when straying from them.

Stresses

In this empirical scenario we are working within, we cannot allow any tensile stresses to develop in our structure. This is a key working design assumption for unreinforced, unstabilized masonry — and it's a really good assumption. This is because regardless of whether or not there is any tensile capacity in the material, there will be load conditions that generate the equivalent of tensile stresses. It will happen, for instance, when the wall is exposed to an out-of-plane bending stress, such as when the wind blows on it from a perpendicular direction.

If we think of the wall as a beam oriented on its end (a common way for engineers to analyze a wall in out-of-plane bending), we can calculate

> **Definition of Stress**
>
> In engineering terms, stress is defined as *a force applied over an area*, e.g. psi (lb-f/in^2) or kPa (kN/m^2).
>
> *Tensile stress* (tension) is applied in such a way that it is trying to pull the material in question apart. Mathematically speaking, tension is applied normal to the plane where our area is defined, directed away from the surface of the plane.
>
> *Compressive stress* (compression) is applied in such a way that it is trying to crush the material in question. Mathematically speaking, compression is applied normal to the plane where our area is defined, directed toward the surface of the plane.
>
> *Shear stress* is applied tangent to the plane, so that it is trying to slice the material in question in half.

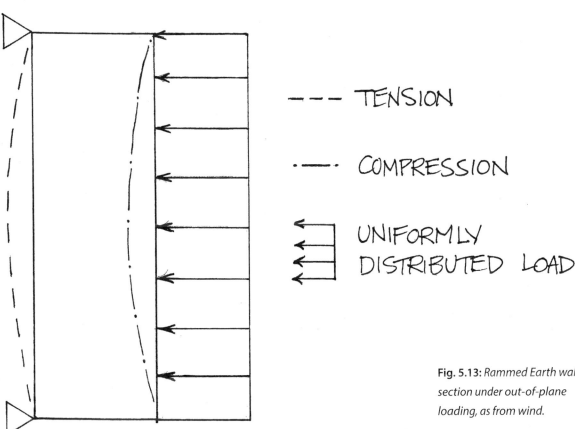

Fig. 5.13: *Rammed Earth wall section under out-of-plane loading, as from wind.*

the bending moment imposed on the center of the wall by a given load. This moment can be expressed as a force couple acting on the wall's cross section; the side facing the load is in compression and the side facing away from the load is in tension. However, we are not allowed to put our unreinforced raw rammed earth wall into tension — so how do we justify this loading (short of adding stabilizer and reinforcing steel)?

We know from experience that unreinforced stone masonry walls are not prone to failure under wind loading, so perhaps this is evidence that a strength-based, limit-states approach of engineering analysis is not appropriate for unreinforced masonry design. That said, a common element of engineering analysis for walls under combined axial and out-of-plane loading is the concept of *eccentricity*. Eccentricity is a way to account for loading at the top of a wall (or column) that is not perfectly vertical and imposed at the center of gravity of the assembly. In the case where a roof element, for instance, exerts a nonvertical thrust at the top of a wall section, eccentricity allows the engineer to create a virtual bending load to account for the horizontal portion of the nonvertical loading.

We are using this concept to limit the stresses in our wall such that the "no tension" condition is more or less guaranteed. Essentially, the

Fig. 5.14: *Sketch illustrating the use of eccentricity to account for nonvertical axial loads.*

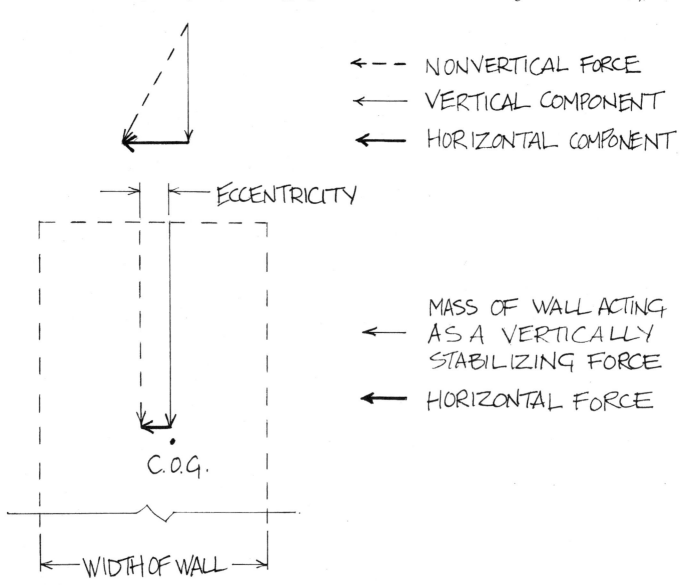

eccentricity created by the lateral portion of the imposed force can be considered to move the vertical portion away from the center of gravity of the wall assembly. In terms of our empirical design guidelines, we do not allow the vertical force to move outside of the walls itself. This is the equivalent of graphically finding the limits of thrust lines within a masonry arch. The MSJC standard limits the application of vertical loads on a wall in this category to the middle third of the wall's thickness. This is an excellent guideline to follow when designing sills or bond beams for anchorage of roof or floor elements.

Wall Length

We know how thick our wall has to be, and how tall we can go, and where we need to apply our loads from above. How long can a wall be? What is the maximum length of wall between lateral supports (which engineers often refer to as a *shear wall*)? The limit depends on the roof or floor diaphragm connected to the top of the wall and the length of the given shear wall. Annex "F" gives us a range of ratios to use based on the stiffness of the diaphragm being proposed. For wood diaphragms, the ratio for maximum spacing between shear walls to the width of floor or roof diaphragm is 2:1.

I suppose it's worth talking a little bit about what a *diaphragm* is, and what we need them to do for us. From a structural point of view, a diaphragm is a horizontally oriented element that can accept lateral loads from the top or bottom of a wall or roof. If we think that the walls of our building make the sides of a box, the diaphragm is the top of the box. If it is properly constructed and connected, the diaphragm will keep the top of the walls straight even when something wants to press the walls inward. Obviously, at the corners of the box, the wall that is perpendicular to the lateral load will keep its partner from tipping over. But what about halfway between,

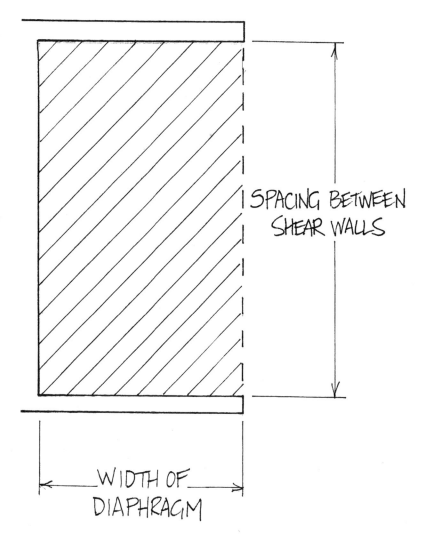

in the middle of the box? It's the diaphragm that will do the work here, keeping the top of the box straight.

Openings and Anchors

There is a certain amount of a wall that needs to remain in order for a wall to be considered adequate to take lateral loading. For a wall with H/T ratio of 14 (the limit of slenderness, discussed at the beginning of this section), 40% of the wall needs to remain. If your geometry has a H/T ratio of 10 or less, you can go down to 20% of the wall.

Anchor bolts for sill plates need to extend a minimum of 150 mm (6") into the top of walls,

Fig. 5.15:
Plan view sketch of diaphragm area above shear walls.

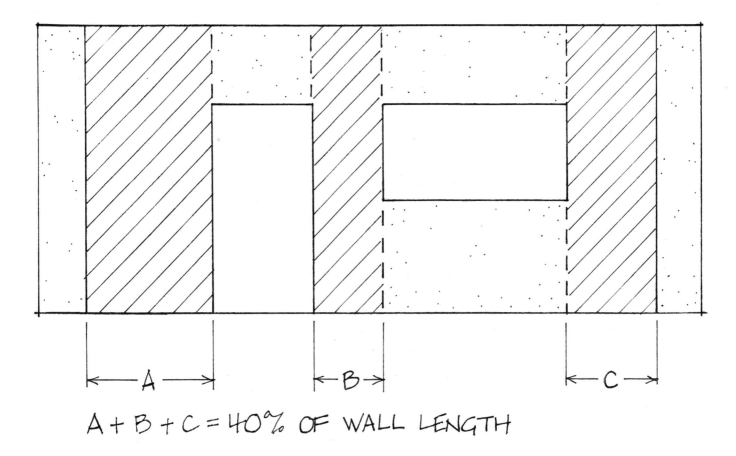

Fig. 5.16: Elevation view of allowable opening geometry in the long exterior wall between shear walls.

Table 5.1: Shear strength for embedded bolts in raw rammed earth

Bolt Diameter	Shear Strength
16 mm (⅝")	1.26 kN (283 lbf)
20 mm (¾")	1.82 kN (409 lbf)
24 mm (1")	3.08 kN (692 lbf)

spaced at a maximum of 1.5 m (5'). If anchor bolts are connecting a roof or other structure subject to uplift, they need to engage a mass of rammed earth with a dead load equal to twice the uplift force.

Other fixings — in particular, embedded bolts or threaded rod — may be necessary to connect non-rammed earth elements such as roof and floor components to the walls. For raw rammed earth, we can take shear capacities for embedded steel bolts from the New Zealand standard. Table 5.1 summarizes the factored capacities of various diameters of bolts.

The required bolt embedment depth is three-quarters the width of the wall element, with a 200 mm (8") minimum. For walls too thin to allow 200 mm (8") embedment, reduce the shear capacity of the bolt by the embedded length in mm divided by 200 mm. For example, a 200 mm (8") wall would have an embedment depth of 150 mm (6"). So, our reduction in shear strength for a 16 mm (⅝") bolt would be 150/200 = 0.75. Applying the reduction factor to the shear strength yields 0.75 × 1.26 kN = 0.945 kN (212 lbf). The NZ 4297 bolt embedment shear strength values for raw rammed earth assumes that the head of the bolt is embedded in the wall, with a 3 mm (⅛") 50 mm × 50 mm (2" × 2") square or 3 mm (⅛") × 55 mm diameter (2¼") round washer in place as well.

The minimum edge distance between the bolt and the edge or end of the wall, measured from the center of the bolt, is equal to the minimum embedment depth. The NZ 4297 standard assumes a minimum compressive strength of 0.5 MPa for raw rammed earth, and advises a minimum wall thickness of 250 mm (10"); however, there is a note stating that 280 mm (11¼") may be necessary to meet minimum thermal requirements without additional insulation. This is well within the range of strengths and geometries we have been working within in our empirical design guidelines.

Lintel Design

The criteria for supporting material above an opening are too varied for me to be able to provide here a simple table of "canned" solutions. Not that I'm looking for more work, but I am going to recommend consulting an engineer for your lintel design, whether building with raw or stabilized rammed earth.

Stabilized Rammed Earth Structural Design

Stabilized rammed earth can be designed using the limit-states provisions of CSA S304.1 in Canada, and following either the working stress or limit-states design provisions in TMS 402/ACI 530 in the United States.

As we know from Chapter 4, Materials, the addition of cementitious stabilizer changes the qualities of rammed earth. In terms of structural design, it becomes stronger in compression, stiffer, more brittle, more resistant to erosion, and better able to engage embedded mechanical reinforcement. This means that we can consider narrower sections for our walls, and longer lintel applications become possible; we can also count on higher resistance to both in-plane shear and out-of-plane bending loads. If you have seismic loading to consider at your location, this increase in strength means that stabilized rammed earth may be your only option.

Durability is an important consideration for engineering design, although it is generally put into a "serviceability" limit state rather than an "ultimate" limit state. For stabilized rammed earth, environmental testing has indicated a minimum (cementitious) binder content of 5% by weight in order to be effective at reducing spalling and erosion due to wet-dry and freeze-thaw cycles. At this level of stabilization, design compressive strengths can reach the 3 to 5 MPa (435 to 725 psi) range. This is certainly a significant increase over the 0.5 to 0.7 MPa (70 to 100 psi) that we are specifying as the minimum acceptable strength for raw rammed earth.

As a masonry design standard, CSA 304.1 allows for unreinforced walls in certain structural situations, depending on geometry and the magnitude of the load. The inclusion of mechanical reinforcement allows for even taller, narrower walls and longer distances between shear wall supports.

It is not the purpose of this book to go through engineering analysis following published design standards. The empirical design guidelines given for raw rammed earth will work for stabilized rammed earth as well. That said, I believe that engaging a structural engineer experienced in rammed earth design is necessary and offers enough value to cover the associated costs when the entire project budget is considered.

An engineering specification for stabilized rammed earth is given in Appendix A. It should be modified to suit the specific project, and may be used for raw rammed earth as well — although the criteria regarding binders would, of course, be irrelevant for raw rammed earth projects.

Design Drawings

I am not a fan of "canned" designs. I believe that consideration of a building's site, the materials being considered, and the lifestyle goals of its occupants are vital to a successful design. These factors, along with cost, labor availability, local regulations, and timelines all need to be weighed properly and as fully as possible before putting pen to paper. Starting with someone else's idea of a "good" floor plan is not the right place to begin. Incorporating the elements of floor plans that you really like is a great idea, but an eye-catching drawing from the internet is not the way to start your building design.

Time spent at the design stage is relatively inexpensive — changes can be made with very little consequence. Once construction starts, however, costs begin to rise — in terms of both money and time — as well as the associated emotional expenses.

Relevant Research

Arrigoni, Alessandro, et al. "Weathering's beneficial effect on waste-stabilized rammed earth: A chemical and microstructural investigation." *Construction and Building Materials* 140, 2017, pp. 157–166.

Delgado, M.C.J. and I.C. Guerrero. "Earth building in Spain." *Construction and Building Materials* 20, 2006, pp. 679–690.

Jayasinghe, C. and N. Kamaladasa. "Compressive strength characteristics of cement stabilized rammed earth walls." *Construction and Building Materials* Vol. 21, 2007, pp. 1971–1976.

Morris, Hugh, Richard Walker, and Thijs Drupsteen. "Observations of the performance of earth buildings following the September 2010 Darfield earthquake." *Bulletin of the New Zealand Society for Earthquake Engineering* 43(4) December, 2010.

Reddy, B.V. Venkatarama and P. Prasanna Kumar. "Structural behavior of story-high cement stabilized rammed-earth walls under compression." *Journal of Materials in Civil Engineering* Vol. 23, March 2011, pp. 240–247.

Walker, Peter J. and Stephen Dobson. "Pullout tests on deformed and plain rebars in cement-stabilized rammed earth." *Journal of Materials in Civil Engineering* (13) 4, July/August, 2001, pp. 291–297.

Chapter 6

Tools and Mixing

Here we're going to go over all of the tools required for both raw and stabilized rammed earth construction. We'll show a few examples of the mix methods that go along with specific types of equipment. More thorough explanations of the formwork and mix placement will be laid out in Chapter 7.

First of all, get organized: It's very important to have a clean(-ish), clutter-free area to make and record binder, additive, and water quantities for each batch. You will probably need to convert between volume and weight-based portions in the recipes, which can be a bit confusing. When dealing with aggregate, especially in bulk, the units will be both volume and weight.

From an engineering point of view, mix recipes are done in terms of percentage of total mass for each ingredient. On paper, this makes a lot of sense; in the actual working environment, weight is a more consistent metric for creating repeat batches of the same mix.

The conversion between volume and mass can be accomplished on paper via density values, but it is much more accurate to use a scale on site to account for daily variations in humidity and the actual density of the given ingredient at any given time. This applies more to the binder and oxide portions of the mix — it is not a simple thing to measure the weight of a skid-steer bucket full of soil mix. Experienced equipment operators are remarkably consistent in scooping up a regular volume of soil mix. Keeping that portion as consistent as possible, and keeping careful track of the binder, admixture, and water portions will make it easier to keep producing quality results.

Fig. 6.1: *Mixing table. Note the two scale types, volumetric tools, and clipboard for recording. White jugs are silicate emulsion admixture and water; blue bags are oxides for coloring.*

Fig. 6.2: *Scale, bucket, and calcined clay for binder mix.*

Fig. 6.3: *"De-mountable" mixing pad with rototiller at the ready.*
Photo credit: Rachel Alpern

Mixing Equipment

The most basic tools for mixing rammed earth are hoes, rakes, and shovels. Mixing can be done either in wheelbarrows or simply on a flat, level area on the edge of the building site. However, working at a scale comparable to modern North American construction requires mechanical mixing.

The first step up from hoes, rakes, and shovels is the use of a rototiller for the mixing muscle. When using a rototiller, you will also need a place to mix that will not result in contamination of the rammed earth mix with organics. Perhaps the simplest way to achieve this is to use concrete sidewalk pavers as the base for a mixing area. Figure 6.3 is a picture of a mixing pad created using pavers, bounded by a wooden frame.

Soil is brought into the mixing area in a known quantity, and the binder and any other dry ingredients are mixed in before adding water, as shown in Figure 6.4. Mix all of the dry ingredients — cement, lime, and any other pozzolans, as thoroughly as possible, as shown in Figure 6.5. Safety should always be of utmost concern: the hazards that come with mixing fine dry powders are often overlooked. Dust masks should *always* be worn when adding/mixing any fine powder — including cementitious binder — to the soil mix.

Fig. 6.4: *Mixing in dry binder using rototiller.*
Photo credit: Rachel Alpern

Fig. 6.5: *Thorough mixing of the dry ingredients before adding water.* Photo credit: Rachel Alpern

Some oxides and pigments are best added to the dry ingredients, and some are better distributed by mixing them in the water portion. Some builders have had success mixing the color in dry, and others find that this results in spotty or blotchy bits of color in the wall — likely due to variations in the concentration of mix water throughout the batch. I recommend following the manufacturer's specification with colors. I also suggest that you make a number of test batches, ideally at close to full scale, before putting a feature lift into your wall.

Figure 6.6 shows a paint mixer and drill set up to mix liquid ingredients before they are added to the total mix.

If you require a higher volume of mix for each batch, larger machinery is required. Some crews have achieved good results using a skid steer to both mix and deliver material. Figure 6.7 is a photo of a skid steer mixing stabilized rammed earth. In this scenario, a pad is prepared on site; usually of concrete at least 5″ thick. The raw materials for each batch are brought to the pad — the soil mix, the binder, the water, and any other additives. An experienced operator then uses the bucket of the skid steer to mix this all together — usually circling around the pile a few times, picking up scoops of material, lifting it and dropping it back down on top of the pile. The mixing starts with the dry ingredients; the dry binder and soil components are mixed as thoroughly as possible before adding the liquid. Remember to limit the volume of each batch to what can be placed and tamped within one hour of the liquid being added to the mix.

Another crew member is ready with water, adding as necessary to reach the optimum moisture content for maximum compaction. It's a bit like kneading bread dough, with a thorough blend being the goal, rather than developing the gluten. Admixtures like silica emulsion sealers and certain colors are best added at this stage.

Fig. 6.6: *Cordless drill and paint mixer to combine color oxides and other admixtures in water before adding to soil mix.*

Fig. 6.7: *Mixing SRE using skid-steer bucket on a concrete mixing pad.*

Fig. 6.8: *Adding color and sealer with the water component of the mix.*

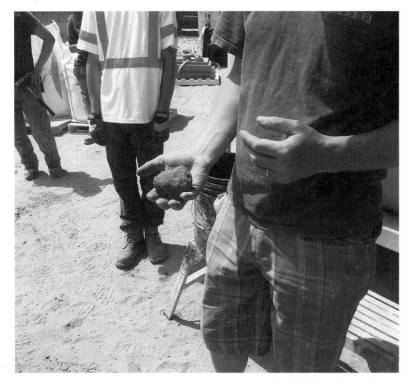

Fig. 6.9: *Ball (not drop) test for stabilized rammed earth batches once the mix design is "dialed in."*

Figure 6.8 shows a crew member adding water with color and sealer to the mix. (Note the use of the face mask.) For the pad/skid-steer method, it is useful to make an indentation in the top/center of the pile to add the liquid component evenly and to make sure it all stays in the mix rather than splashing out onto the pad.

For stabilized rammed earth, this is where the ball test is most useful — to determine when enough water has been added to the mix. Figure 6.9 shows a successful ball test on a work site.

It may be necessary to sieve materials, especially if site soil is being used for all or part of the final mix. Again, at a small scale, a frame with the desired mesh size can be set up over a bin or wheelbarrow and a shovel or buckets can be used to dump the over-sized soil through. It's a good idea to slope the mesh so that you can direct the over-sized material away for use elsewhere. If you are going to do a lot of sieving, you will need to be able to deal with both material streams. In the photo example, the larger material will simply fall in front of the skid steer. A sloped surface directing the material away from the working/driving area would be a good idea.

If your soil doesn't contain lumpy clay, and you are grading out larger pieces of gravel, this can yield relatively "free-draining" material that can be used in other places on the build. Figure 6.10 shows a sieving setup using a skid steer to deliver the soil to be graded.

For crews using dedicated mixing equipment separate from the delivery mechanism, horizontal shaft mixers with flat paddles give the best results for stabilized rammed earth. Mortar mixers can be modified to give consistent results at a volume that keeps several rammers busy. Figure 6.11 shows a stand-alone mortar mixer in use.

Fig. 6.10: Sloped mesh sieving site soil to add to rammed earth mix.

Fig. 6.11: Stand-alone horizontal shaft hydraulic mortar mixer with modified hopper being loaded with earth mix by skid steer.

Loading Equipment

The most basic moving and loading equipment is still just a regular wheelbarrow, now paired with a shovel and/or bucket to get the bulk mix into the formwork. Needless to say, a wheelbarrow and a few shovels are not ideal for large productivity goals. However, for modest projects such as a garden wall, or for making test blocks and practicing tamping techniques, they do the job nicely.

In situations where the mixing is carried out by a skid steer, that same machine usually delivers the material to the forms. Figure 6.12 shows material being delivered into formwork by a skid steer. Figure 6.13 is a photo of workers carrying out the final placement of the mix into the formwork.

If a dedicated mixing machine is mounted on a telehandler, it is used to deliver the mix to the forms. Figures 6.14 and 6.15 show examples of

Fig. 6.12: *Stabilized rammed earth mix being delivered to forms with skid steer.*

Fig. 6.13: *Workers in the formwork use shovels to take stabilized rammed earth mix and place it into the forms evenly and accurately.*

this type of equipment in use. This can be a significant step up in efficiency, although it is site and crew dependent — a large, stand-alone mixer can theoretically service more than one skid steer or telehandler. We'll discuss the various roles and workflow in more detail later, but equipment and tools have a large role to play in worksite efficiency, as well as in the quality and accuracy of the final product.

Fig. 6.14: *Horizontal-axis mixing bucket directly mounted to telehandler. Note the hole in base of bucket for delivery hose. A sliding door that is remotely operated from the cab controls the rate of delivery.*

Fig. 6.15: *Delivering mix to forms with telehandler. Communication between workers inside the forms and the machine operator is critical for safe and effective results. Delivery hose allows workers inside the forms to control where mix is going, but the rate of delivery is controlled by the machine operator. Hand signals or radio communication are important to keep this system working well.*

Tamping Equipment

Hand tampers can be purchased or fabricated with metal, concrete, or wooden heads and wooden, metal, or fiberglass handles. For fine work in the corners of formwork or between densely spaced reinforcing steel, custom hand tampers are often necessary. These can be made using dimensional lumber, such as 2 × 2 or 2 × 4 trimmed to a specific shape. With or without a custom metal head, these hand tampers can be made to fit either angular or round chamfers at corners, or to work around electrical boxes, conduit, or plumbing chases. Figure 6.16 shows manual and pneumatic tampers.

Pneumatic tampers have become the state of the art for rammed earth in most industrialized contexts. Some air tampers are repurposed versions of tools designed to shape large sand molds for iron castings. There are also pneumatic tampers made for various mining operations that can be used for ramming earth inside formwork. You might choose these based on availability, weight, or compatibility with your compressed air setup. But most compressed air systems are compatible with the quick-couplers shown in our example(s). Figure 6.17 shows a builder cleaning and oiling the head of a set of pneumatic tampers. Note that air tools can be finicky, especially when used in the relatively dirty environment of rammed earth formwork.

Generally speaking, the compressed air lines are connected with quick-release couplers that are industry standard. However, the trigger for the tamper is in the upper section, or handle, and the tamper itself is in a separate, lower module. These are connected with threaded pipe — which can be varied in length. Depending on wall geometry and formwork constraints, there may be situations where a worker cannot physically get into a volume. Being able to extend the reach of the tamper is a great option in these cases.

Fig. 6.16a and b: *Manual tampers (left) and a pneumatic tamper (right).*

Air Compressor

If your crew is going to use pneumatic tools for compaction, you will need to be able to supply an adequate amount of compressed air to keep your desired number of tampers going without a significant drop in pressure or volume as each batch is compacted. If more than one person is going to be tamping at once, it is likely that a diesel-powered compressor will be necessary. Because of space limits within formwork, and typical batch sizes, it is common for crews to have two people tamping at once.

The industry standard is the 185 cfm, 80–125 psi, ¾" diameter delivery diesel-powered portable (tow-behind) compressor. One of these units is easily capable of keeping up to eight pneumatic tampers running at 95 to 100 psi. Using ¾"-diameter hoses fitted with common quick-couplers, a typical pneumatic rammer with a 3" to 4" stroke will attain good compaction at this pressure and volume of air.

Smaller, more efficient compressors are available, as are larger units. It is a good idea to work out the details of your ramming equipment when you are preparing test samples during the design phase — unless you have the luxury of access to a setup that has been used for full-scale rammed earth projects in the past.

Fig. 6.17: Regular maintenance: cleaning and oiling of air tools is key to keeping the equipment and crew working on schedule.

Various Necessities

Hoses, pressure regulators, quick-couplers, levels, vacuum, tarps, extension cords, fuel containers, lights, heaters, buckets, paint mixers, and power tools are all needed, as well as some other miscellaneous items.

You will need hoses to connect the compressor to the tamper(s). Quick-connect couplers and a variety of splitters and regulators are available to feed several tampers from a single compressed air source while still delivering consistent pressure and volume to each unit. It is best practice to include an in-line water separator and an oiler in the line between the compressor and the tampers.

It's a good idea to have several spirit levels (aka bubble levels) of various lengths available for setting plumb and level during formwork setup, and to check that levels are maintained during ramming. Another indispensable piece of equipment for setting complicated geometries is a five-point, self-leveling alignment laser level. These units use a mirror system to refract a laser beam and simultaneously project two perpendicular planes and a line normal to both.

Having a vacuum cleaner on site is a must for jobs with any kind of precision as a goal. A wet/dry vacuum is handy for cleaning up detritus, especially at the start of a new day on top of a cold joint, between lifts, on top of insulation, for removing bits of insulation that have been knocked off during ramming, as well as for sucking up pooled rainwater or other foreign objects that have found their way into the formwork.

Tarps and plastic sheets (polyethylene vapor barrier) are indispensable for keeping the mix piles at steady moisture levels in hot or rainy conditions, for protecting the tops of walls, and for providing shade (when necessary).

If you are working in extreme winter conditions (which is not recommended), you may find yourself in need of a high-capacity heating system and insulated tarps in order to meet a tight schedule in cold temperatures. This sort of construction practice has become all too common. I do not recommend this, but I have worked on projects where a large diesel-powered boiler setup heated water that was then piped through the mix piles and the rammed earth in the formwork to keep the material from freezing in a Canadian winter.

Formwork

This is covered in detail in Chapter 7. Suffice it to say here that you will need formwork to create a rammed earth wall. It can be rented or made from framing materials that can then be reused in the rest of the building in partition walls, floor, or roof components. It may even be feasible to resell gently used formply to conventional concrete-forming crews.

Fig. 6.18: Black hoses convey heated water to and from a boiler, orange tarp is an insulated hoarding type.

Chapter 7
Construction Methods

WE HAVE DISCUSSED THE INTRICACIES OF soil mixes, and gone over the basics of preparing a batch of raw or stabilized rammed earth. But we haven't gone into any detail about the formwork that we're putting that mix into.

Formwork

Formwork creates the negative space that the rammed earth fills. The rammed earth will repeat everything the formwork does — as a mirror of the surfaces and a solid casting of the volume created by the forms. Setting up, aligning, bracing, taking down, moving, and setting up the formwork all over again will take considerably more time than the actual ramming of the earth.

Even though the formwork for rammed earth can be compared to concrete forms, the performance of the assembly will be different for several reasons. The formwork has to resist lateral pressure at the height of ramming — it is a dynamic pressure, caused directly by the action of the tampers. When the tamping has stopped, the pressure is largely dissipated, but any deflection outward that is laterally supported by the now-solid rammed earth inside the formwork will remain. As the lifts are added one on top of each other, the pressure exerted at the bottom of the wall system is largely vertical, acting on the foundation. For this reason, raw rammed earth has often historically been built with slip forms that "crawl" up the wall as it is rammed in place, leaving exposed, structurally competent (at least in terms of being able to support its own weight) rammed earth below. This style of rammed earth construction is well documented in other texts, so we will not go into any detail for these *slip-form* techniques. The interested reader can find information from the sources listed in the Bibliography.

The pressure profile and duration on the formwork from the rammed earth is considerably different from cast-in-place concrete that exerts hydrostatic pressure on the forms for several hours until it sets — and days until it reaches a strength sufficient to carry its own weight. The magnitude of the horizontal pressures involved in pneumatically tamped rammed earth is less than the equivalent width of conventionally cast concrete. However, the dynamic nature of the pressures involved in modern rammed earth construction, along with the mass of the earth itself, means that the forms need to be very stiff to maintain a straight, plumb wall during ramming.

Formply

In North America, the most common material making up the surface that the rammed earth is pressed against within the formwork is *formply*, which is plywood that has been coated with some sort of high-density facing. It is possible to use non-formply sheets as rammed earth formwork, but patches and grain patterns on the plywood will transfer through and show on the finished rammed earth. Custom metal forms are also available, but they are not flexible and are easily dented by the tampers.

Both the American Plywood Association (APA) and the Canadian Plywood Association (CANPLY) have created grades of plywood specifically designed for formwork applications. The APA specification is more uniform across both the US and Canada. The CANPLY specification is adopted in a variety of ways by individual manufacturers, each producing plywood to a CSA-performance standard. In both cases, the specification lays out strength, stiffness, and surface finish quality. Some formply is faced with resin-impregnated paper, and some is phenolic treated. In some cases, the edges are factory treated to prevent water uptake into the exposed end grain of the laminations to limit swelling. It is worthwhile to do some research on the products available at your local lumber supply stores. You may be purchasing a large enough quantity of formply to engage a wholesale distributor.

Formply comes in a variety of sizes and number of laminations, either 600 mm (24″) or 1,200 mm (48″) × 2,400 mm (96″ or 8′) or × 2,700 mm (108″ or 9′), or × 3,000 mm (120″ or 10′). There are tolerances for thickness, width, and length, as well as straightness of edges and overall square-ness of each sheet. Purchasing formply in lifts is a way to minimize differences between sheets straight out of the factory, as well as the differences between various manufacturers.

Formply, like lumber, is a commodity that is subject to a fluctuating price based on supply and demand. In my opinion, it is worth connecting with wholesaler distributors to get price and availability information. Note that formply may not be available late in the building season.

As you may expect, higher-grade formply costs more: seven-layer, hardwood-backed fully papered formply is about twice as expensive as entry-level formply. Rammed earth builders in Canada have had more success with fully papered formply than with the phenolic-treated product. The papered formply resists minor dents and scratches, while the phenolic formply is easily damaged by pneumatic tampers and repeated handling. Paper-faced formply is commonly treated with a form release agent during the manufacturing process. Using additional release agent is not necessary, even when building stabilized rammed earth. It is best practice to clean the formply between uses, and a diluted solution of muriatic acid with a sponge mop does a good job of this. Dilute to the manufacturer's specification. Mop it on and rinse with clean water, then allow to air dry. Repeat once, if necessary.

It is virtually impossible to eliminate the outline of any given piece of formply on the rammed earth beneath. That said, care can be taken to align vertical and horizontal joints to minimize this visual aspect. Or it can be highlighted as a feature by intentionally spacing formply at a wider than minimum distance. This is an aesthetic design consideration, but the layout of the formwork supports and the work flow are also affected by these decisions.

If you're careful during formwork assembly and ramming, taking precautions to protect the edges and paper face of the formply, two to three reuses of each sheet of formply as formwork is a reasonable expectation. Depending on finish quality desired, you may be able to continue to reuse beyond this.

If you're only doing one project, you can use the formply as sheathing in roof or floor elements. Other than cutting the formply and making holes with screws, there should be no structural compromise of the material. Considering the minimum thickness I recommend for rammed earth formply is 18 mm ($^{11}/_{16}$″), it will be a more than adequate substitute for conventional plywood or OSB sheathing at 12 mm ($^{7}/_{16}$″). If you are planning on reusing the formply in your building, make sure that any release agents or impregnated chemicals are not going to be an issue for indoor air quality. Sourcing all of your building materials in accordance with the Living Building Challenge's Red List is highly recommended.

Supporting the Formwork

In contrast to many conventional concrete formwork assemblies, where formply may be oriented vertically or horizontally, each piece of formply for a rammed earth wall is set horizontally — with its long axis parallel to the base of the wall. This is because on the side of the formwork where the uncompacted mix is being added during construction, the formply is added sequentially as the wall is rammed from bottom to top, typically in 1,200 mm (4′) sections. The formply is held in place with horizontal components commonly called *walers*. Walers can be made from dimensional lumber, engineered wood, or custom metal sections. Walers are held in place by vertical *strong-backs*. In conventional concrete formwork, the first vertical elements are usually studs, directly placed against the formply. That interferes with the open-sided process that most rammed earth builders use in North America, so studs are generally omitted in this method. Figure 7.1 is a photo of the back side of a section of formwork set up for a relatively short section of wall.

In Figure 7.1, the through-ties shown are made of coil-rod; they are designed to be removed when the formwork is disassembled. Figure 7.2 shows some coil-rod and coil-ties before installation. Note that the coil-rod has a relatively large thread pattern in comparison with conventional threaded rod. Conventional hexagonal nuts are available for tightening with plate washers along with the coil-tie couplers. The removable coil-rod detail is accomplished by installing a coil-tie within the wall assembly, as close to one side of the internal insulation as possible, as shown in Figure 7.3. Figure 7.4 is a photo of the hole left in the rammed earth after the coil-rod is removed. Figure 7.5 is a photo of a blow-out caused by not embedding the coil-tie deeply enough in the wall.

Conventional concrete formwork assemblies often use embedded "snap-ties" — metal ties that are designed to be broken off after the concrete has set, just within the plane of the wall, leaving a length of metal inside and a small divot on the surface of both sides of the wall. Some are designed to have a metal or plastic

Fig. 7.1: *Rented aluminum formwork components, showing base of wall, connection to foundation, through-ties, custom brackets, and high-strength metal structural components.*

convex washer at this location to minimize the damage when the tie is snapped off. In some areas, notably Australia, it is common practice to use these types of ties. For the insulated rammed earth that we are promoting here in our colder climate, this represents a significant thermal bridge and is it best practice to minimize the use of embedded ties and to be careful with their placement.

That said, for interior mass walls or uninsulated exterior walls, such as a garden wall, snap-ties allow for much lighter waler/strong-back combinations. They also make the use of 600 mm (2′) wide formply a practical option. A word of caution: using snap-ties means a lot of horizontally oriented metal inside the formwork that can interfere with worker's mobility. This may

Fig. 7.2: *Lengths of coil-rod and coil-tie couplers. Note hexagonal nut on end of one length of coil-rod.*

Fig. 7.3: *Interior of formwork at base of wall. Coil-tie is installed near the internal insulation, coil-rod is in two pieces, and can be removed from either side separately after ramming is complete and the forms are being stripped.*

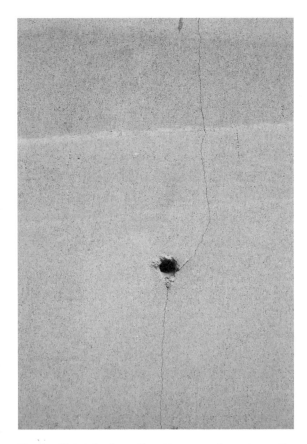

Fig. 7.4: *Hole left after coil-rod is removed near base of rammed earth wall. Diameter is approximately 15 mm (⅝″). Holes can be filled with insulation and capped with pointing grout. Note hair line vertical cracking migrating from footing to hole, then up to a cold joint and further up the wall.* PHOTO CREDIT: DALILA SECKAR

Fig. 7.5: *Blow-out at location where coil-rod was removed. Note location of coil-tie; to prevent this type of blow-out, locate coil-tie further into wall as shown in Figure 7.3.*
Photo credit: Dalila Seckar

not be critical for shorter wall sections where it is possible to access the cavity entirely from the top, outside of the forms. The relatively light waler/strong-back combination may require scaffolding for safe access. I won't go into any further detail about this style of formwork — the interested reader can consult with concrete specialists and use the same principles that I am discussing for wider formply applications.

Cross ties

Most high-energy-efficiency rammed earth builders avoid permanent through-ties altogether, and they carefully choose the locations of removable cross ties such as coil-rod. You can get away with not using through-ties or cross ties by increasing the strength and stiffness of the formwork assembly as a whole and using enough external bracing to push against the sides of the formwork to resist the pressures from ramming. It is important to be very thorough in providing support at the base of the formply in order to keep the bottom of the wall straight and even.

A special case of cross tie is used at the top of the wall, and this is common to all rammed earth formwork systems. It can be a piece of dimensional lumber bolted to the top of each strong-back pair, or a piece of coil-rod or similar high-strength-tension member holding the upper end of the forms together at a set distance.

The APA has a good formwork guide called the "Design/Construction Guide: Concrete Forming." (It is available as a free download or for purchase in hard copy at www.apawood.org/publication-search?q=V345.) I think it is useful to go through the exercise of designing the formwork for a typical pneumatically tamped rammed earth wall section.

Walers

First of all, we need to decide on the thickness and strength of the formply we'll use — which is directly affected by the spacing of our walers. In order to provide even support all the way up the height of the wall, it is good practice to use 400 mm (16″) on center as a spacing dimension. This also means that there is only 200 mm (8″) cantilevered past the center of each waler. Figure 7.6 shows a typical waler/formply layout in section.

Take care to attach the walers to the back of the formply with screws that are not long enough to penetrate the inside face. If the formply is not tight to the waler, the rammed earth will push it out, causing uncontrolled deflection. Screws spaced at 300 mm (12") should keep the waler tight to the plywood. It is a good idea to support the seams between sheets with smaller pieces of formply, as shown in Figure 7.7.

Even with the formply support patches, it is not a good idea to cantilever any edge of the formply more than 200 mm (8") past a support like a waler or end form.

Pressure

Now we have a spacing between walers, but what is the pressure that we're exerting on the formply? As noted earlier, the pressure distribution is mostly vertical, but there is definitely horizontal pressure being exerted on the formwork, and it is delivered dynamically, with considerable vibration working to realign everything not securely fastened. A number for horizontal pressure during pneumatic tamping that is commonly used by rammed earth builders for formwork design is 200 lb-f/ft² (9.6 kPa). In the absence of assembly-specific test data, this is a reasonable number to use.

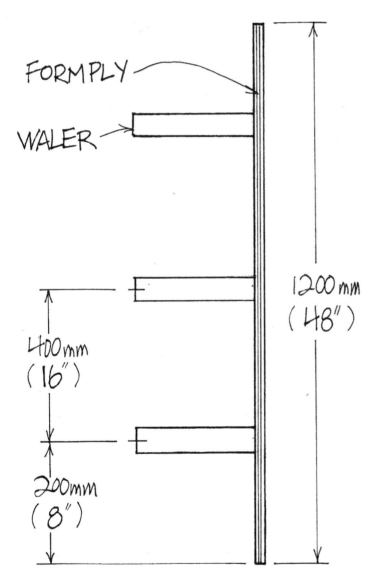

Fig. 7.6: *Walers spaced evenly on a sheet of formply.*

Fig. 7.7: *Formply squares to support vertical seams in formply between walers. Similar application for horizontal seams. See squares distributed along upper waler for installation when next sheet of formply is installed. White strip along top of formply is plastic "J-trim," there to protect the edge of formply during delivery of material and tamping.*

Deflection limit

Another criteria to consider is that of *deflection limit*, which is often expressed as a span over limit — L/360 is an example of a maximum deflection limit commonly used for joists that have drywall attached (where "L" is the span). L/180 is a less restrictive limit, as in cases where there is no ceiling or where something less sensitive to cracking is applied, such as tongue-and-groove milled wood. For example, consider a floor joist with a span of 12'-0" (3.66 m). At an L/360 limit, we have a maximum allowable deflection limit of $^{13}/_{32}$" (10.2 mm). The L/180 deflection limit is twice as much, at $^{13}/_{16}$" (20.3 mm).

Our span between walers is 400 mm (16"). If we want to stay with L/360 as our deflection limit, we can calculate: 400/360 = 1.11 mm (16"/360 = $^{3}/_{64}$"). At a 200 psf (9.6 kPa) pressure, we can choose the thickness of our formply based on expected deflection. Formply that is 19 mm (¾") is adequate for our purposes according to the APA guideline for concrete, but thicker is definitely better, and I recommend using a higher standard for deflection and vibration calculations.

Since our formwork design example is completely made up of dimensional lumber and engineered wood products, we could look to industry standards for direction. The Canadian Wood Council advises a deflection limit of L/600 when designing to account for the effects of vibration in a floor assembly (assumes simply supported joists). Vibration is a factor here, so it may be reasonable to use L/600 as a deflection limit in your design. This means high-grade formply at least 24 mm ($^{29}/_{32}$") thick for our example.

Some builders will see this as overdoing it, especially those accustomed to tolerances typical of many home builders' associations. For instance, in Ontario, the *Tarion standard* allows for 19 mm (¾") out of plumb on a 2,400 mm (8') wall and 15 mm (⅝") from a specified plane for bowing in a wall section. These limits can easily be seen by the eye, whether viewed by an experienced builder or not. To be fair, they are absolute minimums from a document that is designed to determine when it is time to engage a lawyer or make a warranty pay-out. This is not the way I think anyone should design or build. Perhaps a more effective (and economical) deflection limit would be 6 mm (¼") on 2,400 mm (8') for plumb. That works out to L/400.

It is worth noting that if you are going to use metal formwork elements, the stiffness and deflection characteristics of the whole assembly is different from a wood-only setup. Builders in Ontario have had good success with 17.5 mm ($^{11}/_{16}$") high-grade formply on 400 mm (16") centers with aluminum formwork walers as shown in Figures 7.7 and 7.9.

Waler sizing

Now we have our waler spacing and our formply thickness, so what size do our walers need to be? That depends on the spacing of strong-backs, which depends on access points for bringing mix to the forms. For most skid-steer buckets or self-contained mix and delivery equipment, a minimum spacing between strong-backs of 1,800 mm (6') is necessary for safe operation. The 2,400 mm (8') length of formply makes it practical to have offset spacing of strong-backs — or to think of them as being installed in pairs.

In our example, at 600 mm (24") spacing between walers and 1,800 mm (6') spacing between strong-backs, we need to use a minimum of 2 × 12 SPF No 1/2 dimension lumber for the walers. Supporting the walers is a little bit tricky — they need to be bearing against at least two strong-backs to be effective. Figure 7.10 shows an example of NOT properly supporting the walers. This results in more movement in the formply and a wavier surface on the wall, so

88 Essential RAMMED EARTH CONSTRUCTION

it is not recommended. The builder chose to do this in order to avoid cutting any of the waler material (and likely because they believed it was faster). With good planning, it can be done such that each waler is supported by at least two strong-backs, and cutting can be minimized to preserve the dimensional lumber for use in future framing within the building.

Strong-back sizing

We have our walers sized, now how about the strong-backs? It depends on how we are

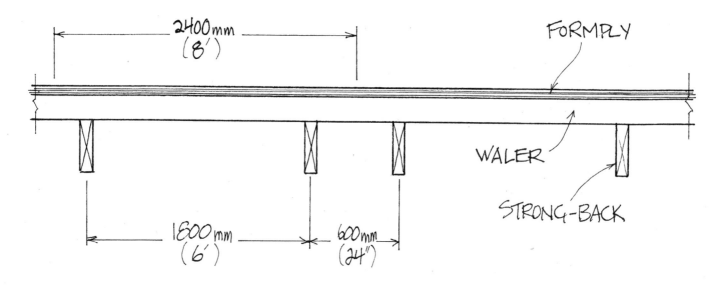

Fig. 7.8: *Plan view of strong-back layout on delivery side of formwork assembly.*

Fig. 7.9: *Formwork needs to be spaced to accommodate delivery of material and workers. In this example, high-strength aluminum rental formwork structure allows wider spans between strong-backs.*

Fig. 7.10: Walers not supported by more than one strong-back. Note that the builder did provide bracing for the strong-backs in both directions, as well as shade for the workers and the mix in the wall to prevent rapid drying out of the mix.

connecting and bracing them — if we consider a 3,600 mm (12′) tall set of formwork, connected at the bottom and top only — as shown in Figure 7.9, where rented commercial aluminum form structures are being used — we need to use an engineered wood product. Two 1.75″ × 9.25″ 2.0E 4620Fb laminated veneer lumber (LVL) pieces will do the job — and just meet the L/600 deflection limit laid out above.

If we brace the strong-backs, as shown in Figure 7.10, at least once in each principal direction at about mid-height, we can use a single LVL of the same dimension and the deflection goes all the way down to L/1440 (1.25 mm or $^{3}/_{64}$″). This is a strong argument for using bracing liberally on your formwork. However, more bracing places restrictions on access, so a balance is necessary to have an efficient work flow that doesn't create frustration for machine operators or workers in the forms.

If you are using wood strong-backs, a useful technique for keeping walers in line is to incorporate metal angle shelf brackets at the appropriate spacing, as shown in Figure 7.11.

Fig. 7.11: LVL strong-backs with metal shelf brackets at appropriate waler spacing. Note also connection across the top of strong-backs above and the wall section end form panel in the foreground. This end panel uses dimensional lumber instead of plywood to accommodate connections where the insulation needs to continue past the vertical cold joint. The middle board is removed in those cases and the insulation extends 38 mm (1½″) past for continuity in the air barrier.

Insulation

Assuming that you are building an internally insulated wall section, it is best to start with a 300 mm (12″) strip of insulation, followed by 600 mm (24″) sections after that, so that the insulation is always a minimum of 300 mm (12″) below the top of the formply, with a maximum of 900 mm (3′) at the beginning of each 1,200 mm (4′) section. This makes delivery of materials into the formwork a lot easier, so it is worth the time and effort it takes to rip down the insulation. Figure 7.12 is a photo of the interior of formwork ready for mix delivery. There is a volume displacement strip on the interior side to make an indentation where the floor slab will be cast. There is a thermal break high-strength rigid insulation base on the exterior wythe, and spacers are in place to hold the insulation in the plane of the wall where we want it. So, in this example, we have a 200 mm (8″) external wythe, a 150 mm (6″) internal wythe, and 200 mm of mineral wool insulation between. The vertical reinforcing steel has been drilled and epoxied into the footing below the wall.

This method is based on insulation materials that are manufactured in 600 mm (24″) widths, as most rigid foams and mineral wools are. Some rigid insulation products come in 1,200 mm (48″) wide sheets. These will need to be ripped down to accommodate the method described here.

Mycelium-based rigid foams are becoming available, and they hold great promise for use in rammed earth wall systems. Fiberboard products made from cellulose are also an option — although they have a relatively low insulation value per thickness when compared to other products. I have always wanted to work on a project that combined straw bale insulation with rammed earth, but straw is not likely to be practical as internal insulation. It could be a good option for externally applied insulation. Paired with a raw rammed earth wall, it would be a low-embodied-energy carbon-sequestering assembly — albeit a very wide one. Keep in mind that such nonconventional insulation materials may not be available in convenient dimensions to match the regular formply sizes.

End panels

End panels need to be solid in terms of deflection and vibration, but they may also need to accommodate embedments for attaching windows or doors, or facilitate the continuity of an internal insulation layer. The simplest end forms rely on the tensile capacity of the formply and a solid connection to each side of the rest of the formwork rather than an external bracing

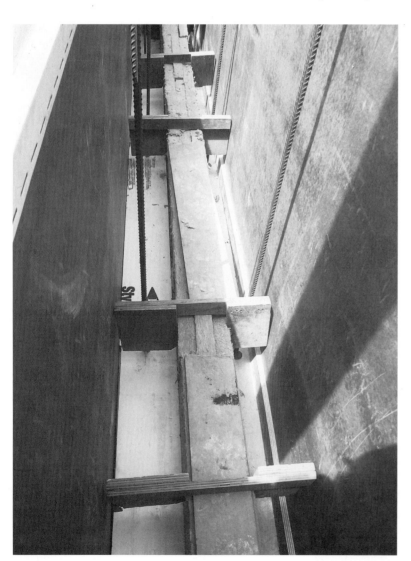

Fig. 7.12: Interior of insulated rammed earth formwork ready for mix delivery.

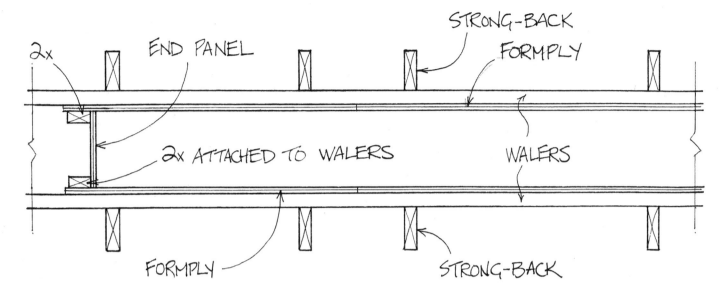

Fig. 7.13: Plan view of end panel in formwork assembly.

system. The end panel needs to be the full width of wall section, two layers of plywood thick, with 2× material to connect to walers. Screw the long-wall ply into the 2× to avoid pillowing (on 400 mm or 16" o/c between walers — making a positive connection every 200 mm, or 8"). Figure 7.13 is a sketch of an end panel in plan view, installed in a formwork assembly similar to the one in our design example.

Avoid adding too many screws in positions where you cannot remove them later — it can make taking down the formwork very difficult. Considering the disassembly process can be just as important as thinking about the structural integrity of the formwork.

Pipe clamps can be used at the top and ends of smaller wall sections; but be careful because they tend to slip, and they do not have capacity to handle larger loads. Threaded-style rod is more dependable. The coil-rod product mentioned earlier is a great option for this — it can be embedded and removed and used at any height of the wall to keep forms parallel. Coupler nuts and other accessories for set angles other than 90° are available. Coil-ties can be used to anchor embedments and even open up the possibility of one-sided formwork systems.

The choice of form system and work flow strategy has implications on the aesthetics of the final building. Some builders have refined systems to make freestanding straight or corner sections and then either use wood framing or set up more formwork between the completed rammed earth sections to add under-fill and/or over-fill rammed earth at windows or doors. This method eliminates long continuous sections of rammed earth wall, and can make for relative ease of access around the site. It also results in many vertical cold joints. It is a challenge to make the individual wall sections line up, potentially creating finishing and cabinetry problems for later on in the project. Figure 7.14 is a photo taken at the infill stage of a project built using individual sections. Figure 7.15 is a close-up photo of one of the infill sections, showing the insulation layer and connection to the previously completed rammed earth wall.

A major advantage of this method for the builder is the ability to reuse formwork several times in a single project. Therefore, much less material is required, in contrast to a system where all the walls are formed at a single time.

End panels may also have to accommodate different wall types and maintain continuity of

Fig. 7.14: Under-fill sections being installed between separately constructed wall sections. Note the tarps at the ready to cover both formwork and walls in case of rain.

Fig. 7.15: Close-up of infill section between previously completed rammed earth walls. Note the manual tamper on the waler at the right.

insulation, as shown in Figure 7.17. The end panel on the right side of the photo marks the end of the rammed earth and allows for a framed wall to connect at 90°, while the insulation layer stays continuous. This end panel features chamfered corners. The end panel on the left side of the photo is meant to connect with more rammed earth wall along the same line — the insulation extends through to prevent air leakage in the cold joint formed by this arrangement.

This photo also shows a section of wall where formply pieces were not lined up consistently and different sizes of plywood were used. This may not be a critical element in your design, but if it is, it must be accounted for in the initial formwork construction.

Outside Corners

You can minimize the sheets that need to be cut by thinking about which side you allow to run long. Caulking the corner makes using non-factory edges/cut lines a moot point. If you are overlapping walers to brace outside corners, you can't run one sheet of ply long, so you will have to cut some of your sheets.

Construction Methods 93

Fig. 7.16: *Rammed earth house under construction; three larger sections of wall are visible, with one infill section between two of them. (Dark vertical lines are from rain earlier in the day and are not defects.)*

Fig. 7.17: *3,650 mm (12') high wall section with one 90° corner. Note chamfered corners and side-exposed insulation at near end. Insulation is exposed at the far end; this wall will continue in a straight line. Also notice the formply lines are not lined up evenly — they are offset on the top layer, and the builder used pieces of different sizes. The results are visible.*

Fig. 7.18: Formply connection detail for inside of 90° corner.

Getting outside corners plumb is not always easy. You can tie 45° strong-back pairs together with coil-rod, and this may be able to hold the corner, but setting them plumb over the full height can be problematic because you can only tighten (not loosen) the coil-rod. Tightening does not always move the side that you want moved. If one side is perfectly plumb and you tighten, trying to use the plumb face as reference, it may come out of level rather than pulling the other side to where you want it.

Patience is critical if you want a very straight, level wall. This may seem pedantic, but anyone who has tried to install kitchen cabinets against non-plumb, non-level, non-straight/flat walls can tell you that it's worth getting things as straight as possible. Compound errors over a 3,650 mm (12′) or 4,570 mm (15′) wall can easily add up enough to affect the location of roof bearing. It is a good idea to check back to the base of the wall with a multi-point plumb laser *at every formply row* to avoid compounding errors in plumb, level, and alignment along the line of the wall.

In many cases, roof components such as pre-engineered truss packages need to be ordered well in advance of the completion of the walls. It is costly in terms of both time and money to have to modify other building components in order to accommodate inaccurately built rammed earth walls.

It is not too difficult to make an outside 90° corner sharp, but the inside corner is more of a challenge. If you use a basic butt joint, you will see the end grain of one piece of the formply embedded in the finished rammed earth. Figure 7.18 is a sketch of one possible inside corner

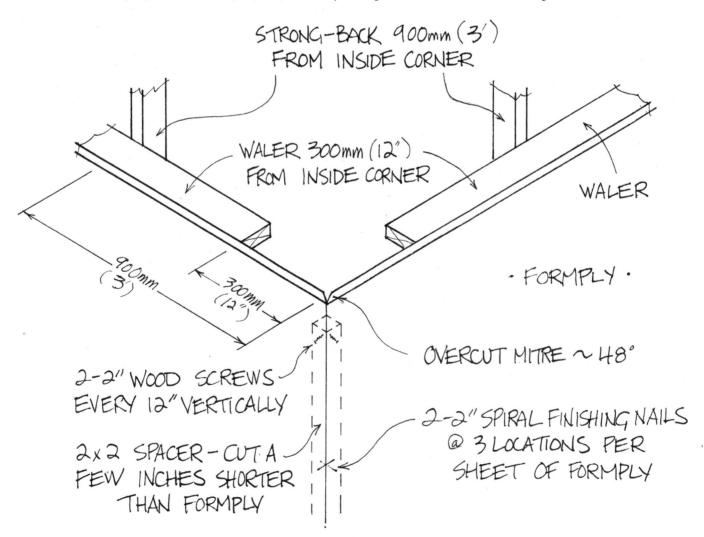

assembly, designed to avoid any formply end grain appearing in the rammed earth.

To create clean, slightly rounded corners and to minimize chipping, it is a good idea to use silicone caulking along vertical joints at both inside and outside corners.

If creating sharp 90° corners is not desirable or not easy (another reason to build some test walls ahead of time) — the addition of a chamfer in the corners of the formwork is one solution. It is not that easy to rip 8'-lengths accurately with a circular saw or job site table saw. Pre-cutting the infill material or getting a cabinet shop to provide them are both good options.

Volume Displacement Blocks

If your aesthetic includes long, continuous lift lines, you will need to use longer formwork; rental materials may become the more economical choice in this case. This may change the way you choose to put in openings, favoring Volume Displacement Blocks (VDBs), rather than infills between previously rammed earth sections.

VDBs need to be strong enough to support the weight of workers and the first several lifts of mix above, or an embedded lintel if that is being incorporated. Figure 7.19 shows a VDB being constructed from formply and dimensional lumber. In this case, the bottom of the opening has been cast with a wet mix and given a troweled finish to the desired base elevation of the opening. A piece of formply has been added to the inside of the formwork to support and locate the VDB, and to create a sharp line along the wet mix layer. Figure 7.20 shows the troweled finish and the formply supports as a builder vacuums the area before the rest of the VDB is installed. Figure 7.21 shows the top and sides of the VDB being put into place.

It is important to mark the location of VDBs on the exterior of the formwork assembly by spraying highlights or marking paint on screws

Fig. 7.19: *Volume Displacement Block (VDB) being assembled; this design is three-sided to join with previously set and cast bottom in the formwork.*

Fig. 7.20: *Trowel finished bottom of opening, with formply strips on each side of formwork to accept rest of VDB frame. Note vertical reinforcing steel on side of opening and builder cleaning the area with vacuum before installing the rest of the VDB.*

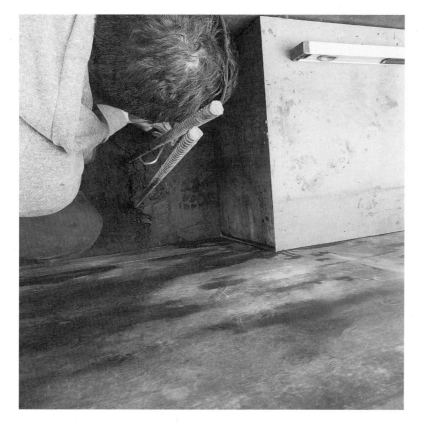

holding the VDB frame in place. Be sure to remove each of these screws before stripping forms. Otherwise, when you attempt to take the formply off the RE, you will ruin the edges of the void you so painstaking tried to create. Figure 7.22 is a photo of the opening created in this example.

Screws and Nails

Screws are the most common connectors used when working with the formply and elements like VDBs, electrical boxes, etc. This is for ease of disassembly and reuse of formwork elements. Screws are good at resisting pull-out, so they can be used where tension resistance is required. High-strength screws do have good shear resistance, but depending on the thread pattern and location, they may not be as effective as nails. Double-headed nails are commonly used in formwork assembly because they are easy to pull, and they provide a very good shear connection.

Fig. 7.21: *Checking level of upper VDB frame, caulking joints between formply surfaces.*

Fig. 7.22: *Formply removed and interior VDB frame still in place. Note the length of wall: this is 48.75 m (160') of continuously formed stabilized rammed earth, combining insulated and uninsulated sections.*

PHOTO CREDIT: JAMES BLACKMAN

General Tips and Techniques

Builders should aim to use all of each batch of mix in a single lift, staying between 150 mm (6″) and 300 mm (12″) loose depth. Lifts can be tapered down to a point if you are building longer walls than a single batch can cover. In general, you should see a two-third reduction in depth by ramming — i.e. 150 mm (6″) loose will become 100 mm (4″) compacted. There are diminishing returns as you go thicker, and 250 mm (10″) is the maximum recommended compacted thickness. More is not advised. Narrower lifts (above the minimum necessary for the tampers to be effective) are always stronger and more consistent. Staying with narrower lifts will also help keep color consistent, especially if you have oxides and pigments in your mix.

Have a piece of 10M rebar with tape at the range of lift depth handy to make sure you aren't delivering too much or too little material in any given area.

Lifts on top of cold joints — i.e. the first lift of the day above yesterday's/last week's work — should be a bit thicker, rather than thinner. Depending on how tight the formwork is at the edges of the existing tamped material, the top of the existing rammed earth could be wetted down to dampen it after vacuuming off any debris, but before adding fresh material. The wetting should be carried out for raw rammed earth. However, with cement-stabilized rammed earth, this might change the wall's appearance. If the formply has pulled away from the rammed earth (or the rammed earth has shrunk a bit horizontally as it begins to cure), the added water can run down the form face, streaking the finished surface. This is not a structural issue, just an aesthetic consideration.

If you have high air pressure and smaller ramming heads on your tampers, the heads of the tampers may dive down into the material at the beginning of each lift. If this is happening above an older lift, material at the top of the old lift can chip and break. This can be at least partly mitigated by walking back and forth along the top of the loose material a few times before ramming to help keep the tampers from diving too deeply into the lift. Figure 7.23 is a photo of the top of a layer chipping when a thin lift of highly pigmented rammed earth was installed directly above.

It is important to practice banking the loose material off the interior side of formwork with spades while transferring the mix into the forms. This will keep finer material at the exterior and cause larger aggregate to bounce back toward the middle of the wall/wythe. Throw the material

Fig. 7.23: *Chipped edge of lighter-colored rammed earth directly below dark lift. Care must be taken when installing thin lifts. Note also the bleeding of water around the heavily pigmented lift; the builder forgot to add the silicate emulsion internal sealer to the tinted batch, and this material is responding to precipitation differently than the material surrounding it.*

Photo credit: Dalila Seckar

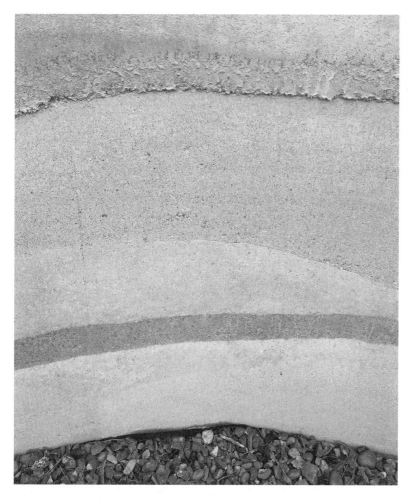

Fig. 7.24: Different colors used in various lifts. The darker tinted lift at the bottom of the photo was placed with a correct amount of water in the mix; the upper, lighter colored lift shows the effects of excess water in the mix.

at the exposed face by flicking loaded spades at the formply, toward the side of the formwork creating the surface that will be exposed.

Experienced builders will often talk about the sound of the earth "ringing" when it is fully compacted — this is also true of manually tamped earth, although to a lesser extent. It is possible to over-compact a lift, but if you have your moisture content correct, you will hear a distinct change in the tone of the tamping head striking the earth when it is compacted to its maximum density.

Safety While Working Inside Forms

Most North American safety standards prohibit having workers in any kind of enclosed space more than 1.5 m (5′) deep without the use of structural cages to prevent injury from collapse of either an excavation or unfinished construction materials. The limits of this depth may be reached for short periods of time when formwork is added — for instance, when the upper surface of rammed earth is 300 mm (12″) below the top of the lower sheet of formply, when the next 1,200 mm (4′) sheet is added. Until the insulation is put in and jigs are put on to allow stepping on the insulation — a worker may be considered to be 1.5 m (5′) down. It is advisable to check with local safety regulations and always conform to best practices for personal protective equipment (PPE), rigging, and tying off when working at heights. Aside from the "regular" PPE (steel-toed boots, safety glasses, gloves), it is important to protect your lungs when working with rammed earth. Dust masks are recommended during all mixing, delivering, and ramming processes.

Rammed earth builders in California have developed a system using deep form sections where the workers are always outside the formwork. They deliver material to the depth required using a 150 mm (6″) diameter flexible pipe (known as an *elephant's trunk*). The pipe is equipped with a 19 mm (¾″) bolt across the end to cause the larger grained material to stay away from the surfaces of the formply. This fulfills the same purpose as banking the material off the side of the forms manually using spades.

In the deep form method, pipe extensions need to be used and interchanged in order to get the tamper heads down to the elevation of the loose material. David Easton's crews have regularly rammed walls 3 m (10′) tall (and some even taller) using this technique.

With insulated, twin-wythe rammed earth, having to place insulation in 600 mm (2′) sections (and horizontal rebar and volume displacement blocks, electrical boxes, and other utility chases) pretty much means having to have a worker in the formwork. This is always easier and safer using the staged formwork system outlined above.

Work flow

There are three main crews: one mixes and delivers the rammed earth to the forms (and makes test samples, when appropriate); another crew preps the next section of formwork (installing insulation, VDBs, electrical boxes, cleaning, etc.); and the third crew distributes material into the forms and does the ramming. It's good to have two sections of wall available at a time that can handle a batch of mix per lift. This way, one part of the crew can be getting ready while the other rams. Depending on the complexity of the build, this means at least six workers; eight is often better.

If you are building stabilized rammed earth, it is very important to make sure that each batch is delivered and tamped within one hour of the binder being added to the mix. So the work flow should be optimized with this in mind: keeping the mixer busy means that the tampers have to be working if all of the mix is going to be used in a timely fashion.

If you are building with raw rammed earth, maintaining the optimum moisture content for compaction is still very important, but there is no limit to the amount of times that water can be added and the batches re-mixed if they dry out.

Removing formply

For raw rammed earth, it is best to use the *slide method* of removal to maintain a smooth wall surface. This means sliding each sheet upward along the plane of the wall surface until it releases from the rammed earth; only *then* do you tip the sheet back and away. For stabilized rammed earth, tip the formply back right away and be careful *not* to slide it along the wall surface. This is because loose bits of material will inevitably fall in between wall and formply, so sliding the formply along the wall will scratch the surface.

A Cautionary Tale

A recent project that made use of volunteer labor ran into a couple of issues that are worth mentioning as a cautionary tale.

First of all, the owner-builder underestimated the structural requirements for the formwork itself, using smaller sections of lumber and thinner plywood than was recommended in the design example above. This meant that the wall bulged out as it was being rammed, creating an effect similar to lap siding — albeit not in an extreme manner, and not with a disastrous result. Some practitioners may actually purposely build for this effect. Figure 7.25 is a photo of the lifts bulging out over each other.

Fig. 7.25: *Note bottom of each lift hanging over the top of the lift below. This is the consequence of the formwork structure not being stiff or strong enough to resist the vibration and pressure of the rammed earth process.*
PHOTO CREDIT: RACHEL ALPERN

The owner-builder had arranged to have another builder who had some experience building rammed earth walls come and train and supervise the volunteer crew. While the supervisor was present, material delivery and lift depth was maintained consistently. However, the supervisor didn't stay on site for the entire build. When the volunteers out-numbered the experienced crew members, the lift depths increased dramatically, and the lower portion of each lift was not adequately compacted.

However, the bulging didn't really become obvious until the forms began to deflect visibly on the outsides. The full effect of the unsupervised volunteer labor was revealed when the forms were stripped. The exterior side of the wall needed manual parging by trowel before backfilling. Figure 7.26 is a photo of the finished wall on the interior side.

The moral of the story is to err on the side of caution with the structure of your formwork, and to be aware that too many unexperienced but eager workers can do more harm than good if they are not kept under watchful eyes. A rammed earth wall, particularly a stabilized one, is a very durable thing. A few hours or even days spent taking care in the formwork is not very much time in contrast to 100 or even several hundred years of service.

Fig. 7.26: *Lifts in the middle section of this wall are too thick. Lower section lifts were mixed and tamped under supervision by an experienced rammed earth builder. Thick middle lifts were done with more volunteers on site than supervisors. Upper lifts were done with more care. Thickness goes down again.* Photo credit: Rachel Alpern

Chapter 8

Cost Estimates Based on a Cement-Stabilized Rammed Earth Project

THE BIGGEST COST COMPONENT of rammed earth walls is the labor, even when equipment such as pneumatic tampers, large-scale mixers, and telehandlers are used. Add to that the fact that exterior walls are not usually the largest cost component of a standard building project, and it is unlikely that rammed earth is going to be a low-cost option in the North American construction market. The percentage of total budget taken up by the exterior walls will depend on the complexity and the size of the design, which includes a host of factors: finishes, mechanical systems, openings for windows and doors, etc.

That said, in this chapter I will present estimates for the four largest costs associated with rammed earth walls: *materials, design, equipment rental,* and *labor.* Figure 8.1 is a pie chart of the cost breakdown for the walls in a small, single-story stabilized rammed earth house project completed in 2015 in Ontario. Table 8.1 gives the dollar amounts as well as the percentage of the total wall cost for each category.

The **Materials** category includes aggregate, the binder mix, color, internal sealer (Plasticure is an example), reinforcing steel, electrical boxes and conduit, insulation, fasteners and miscellaneous hardware, and formply. Figure 8.2 is a pie chart with the breakdown of these costs, and Table 8.2 is a summary of the dollar costs and percentage of the category cost for each component.

The binder for this project was 60% Portland cement and 40% Metapor (metakaolin calcined clay product from Poraver). Internal sealer was Plasticure by Tech-Dry and is included in the binder cost category. The aggregate blend was a combination of sand, crushed gravel, sieved pit run, and crusher fines, which were all delivered to the site. The insulation was 150 mm (6″) of rigid polyisocyanurate.

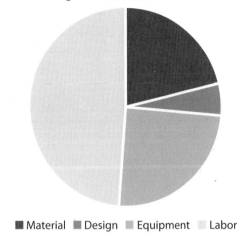

Fig. 8.1: *Cost categories for a stabilized rammed earth example project.*

Table 8.1

Category	Cost	Percentage
Material	$11,825.00 ($14,800.00 CD)	21%
Design	$2,876.00 ($3,600.00 CD)	5%
Equipment	$13,823.00 ($17,300.00 CD)	25%
Labor	$27,166.00 ($34,000.00 CD)	49%

Note: The first figures given are in US dollars.

The per metric ton (1,000 kg, or 2,200 lb) cost of the mix breaks down as follows:

- Binder: $22 ($28 CD) (Portland cement $14 [$18 CD]; Metapor $8 [$10 CD])
- Color: $8 ($10 CD)
- Plasticure: $7 ($9 CD)
- Aggregate material: $14 ($18 CD)
- Aggregate delivery: $8 ($10 CD)

Since the aggregate portion of the mix doesn't have a high influence on the cost but is the bulk of the volume of a rammed earth wall, cost estimates are usually carried out per unit area of vertical wall surface. Builders will use the expression *face feet,* meaning a square foot of vertical wall area. For the 135 m^2 (1,450 ft^2) single-family residence, there were 40 metric tons of mix in the insulated walls. Materials were sourced for 40 metric tons of total mix, including allowances for waste and other contingencies. The project had a total of 1,170 face feet (110 m^2) of stabilized rammed earth, and cost an average of $47 ($59 CD) per face foot ($514 [$643 CD] per face meter).

The **Design** category includes engineering and sealed (stamped) structural drawings for permits and construction. It should be noted that this project used a "known" mix and a simple wall geometry that didn't require any extra laboratory testing; only a minimum of professional design services were needed. Unless you have a similar situation, you should budget at least 5% for the design portion of the wall costs, including an allowance for testing. Add more if the project is exceptionally complex or complicated.

The **Equipment** category includes forming materials (rental of walers and strong-backs), as well as mechanical equipment (telehandler, skid steer, mixer, compressor, diesel fuel, and mobilization). Figure 8.3 is a pie chart illustrating this cost breakdown. Table 8.3 is a numerical summary of the equipment costs and their percentages of that category's total.

The **Labor** category is just that — the person-hours required to do the build. This project was carried out by a crew of eight. Planning for work flow is important for budgeting and for construction. More efficient use of labor will help control costs, since labor is one of the largest costs.

At the time of writing, the price range per face foot of wall for stabilized, insulated rammed earth in Canada was $119 ($150 CD) at the high end and $48 ($60 CD) at the low end (which comes to $1,278 [$1,600 CD] to $511 [$640] per meter). The upper number includes some profit for the builder, travel, and accommodation for a skilled crew. The project itself had a complex geometry with minimal economies of scale. The lower number would be for a simpler geometry and a smallish scale — 140 m^2 (1,500 ft^2) or slightly larger, but single story,

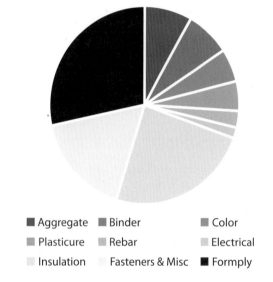

Fig. 8:2
Graphical breakdown of Material category costs.

Table 8.2: Material categories table

Component	Cost	Percentage
Aggregate	$959.00 ($1,200.00 CD)	8%
Binder	$895.00 ($1,120.00 CD)	8%
Color	$639.00 ($800.00 CD)	5%
Plasticure	$575.00 ($720.00 CD)	5%
Rebar	$319.00 ($400.00 CD)	3%
Electrical	$208.00 ($260.00 CD)	2%
Insulation	$2,976.00 ($3,600.00 CD)	24%
Fasteners & Misc	$1,997.00 ($2,500.00 CD)	17%
Formply	$3,356.00 ($4,200.00 CD)	28%

Note: The first figures given are in US dollars.

and no tall walls. These numbers are based on similar breakdowns to the example given, and they include materials, equipment rental, and labor for the walls themselves.

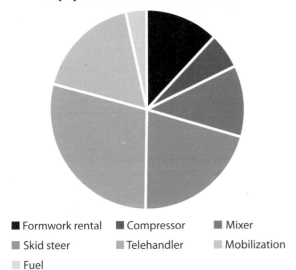

Table 8.3: Equipment cost breakdown

Component	Cost	Percentage
Formwork rental	$1,678.00 ($2,100.00 CD)	12%
Compressor	$799.00 ($1,000.00 CD)	6%
Mixer	$1,598.00 ($2,000.00 CD)	12%
Skid steer	$2,876.00 ($3,600.00 CD)	21%
Telehandler	$3,995.00 ($5,000.00 CD)	29%
Mobilization	$2,397.00 ($3,000.00 CD)	17%
Fuel	$479.00 ($600.00 CD)	3%

Note: The first figures given are in US dollars.

Fig. 8.3: *Graphical breakdown of equipment costs.*

Chapter 9

Wall Surfaces, Openings, and Embedments

AESTHETICS ARE SUBJECTIVE — but rammed earth's aesthetics are the first thing that draws people in, at least here in North America. We covered a lot of process options that have finish implications in the formwork examples back in Chapter 7. Intentionally and successfully achieving the color, texture, surface finish (either smoothness or roughness), and geometry of wall that you want means doing a lot of planning, testing, and preparation. Some rammed earth builders in Canada have taken to using sandpaper "grit" numbers as a measure of finish smoothness. Do you want 60 grit wall surfaces? 600 grit? Somewhere in between? Do you want intentionally exposed aggregate? Embedments? Intentional patterns on the formwork?

We've already covered decisions about the location and geometry of form lines and the impression that the form material will leave behind on the rammed earth. We've gone over material selection, admixtures, and how to build in successively rammed lifts. This chapter will deal with installation of utilities like electrical outlets and conduit and windows and doors, and several design and finish options will be discussed.

Window and Door Openings

This may seem obvious, but rectangular openings in walls have a distinct top, bottom, and two sides. How your openings are created should be informed by how you need these three (and possibly four, if the two sides are different) scenarios to connect with the window or door unit you are using to occupy that opening.

What is the mounting system for your window/door? Most residential window and door frames are designed to work with light wood framing, at least here in North America. But there are also units that are made to work with masonry, and some have adaptable attachment systems.

It may be necessary for you to coordinate between different door and window manufacturer's frame designs if you are using more than one product style in your project. Be aware of the door and window mounting requirements before deciding on your formwork strategy, and incorporate your insulation and finish design into the entire build. I also recommend thinking about repair and future replacements, since it is very likely that the walls will last many years longer than each window and door unit.

Some windows are connected to the rest of the building via a nail flange that can be present on the sides and top only, or on all four sides of each unit. These require a nailing surface parallel to the exterior of the wall. Figure 9.1 is a plan view sketch of one possible way to anchor the side of a window with a nailing flange to an insulated rammed earth wall assembly. Some builders use recycled plastic decking product as a substitute for wood as the embedded nailer material. This style of mounting also works for more traditional windows that have a brickmold trim.

This detail is relevant for the sides and top of the opening, but it requires some modification to work on the bottom. Figure 9.2 is a section showing one option for the bottom side. In this design, a separate sill needs to be installed on both the exterior and interior of the window unit. Pre-cast concrete and stone sills are commonly available for use with veneer masonry

106 *Essential* RAMMED EARTH CONSTRUCTION

exteriors. Figure 9.3 is a photo of an exterior side pre-cast sill installation. Another option is to incorporate custom metal flashing to create the water barrier on the exterior side, as shown in Figure 9.4.

It is not obvious in Figure 9.4, but this window is below an extensive overhang, so the metal flashing here is adequate. For a more exposed window location, I would advise having the finish material extend past the vertical surface of the wall below the window opening and incorporate a drip edge to keep water off the wall and protect the interior insulation from potential leaks.

Figure 9.4 is an example of using rammed earth as the final horizontal surface in the window opening. As you can see, it is not perfectly level. It is very difficult to achieve a level, smooth surface in rammed earth compacted with pneumatic tampers. The installation in Figure 9.3, with the cast sill, incorporates a mortar bed above the top of the exterior rammed earth wythe to set the sill. This is a good detail for durability and air sealing at the base of the window opening.

Another option is to use a high-compressive-strength insulation like foam glass or autoclaved aerated concrete to bridge the internal insulation layer at the bottom of the opening. These types of insulation come in masonry-like units and are designed to be set in mortar. Foam glass can be feathered down to accept exterior steel trim. Or the insulation "bricks" can butt against a separate stone/concrete sill of the type shown in Figure 9.3. Alternatively, you could mortar down a thin stone/tile sill on top of the insulation.

Fig. 9.1: *Plan view sketch of nailer material embedded into rammed earth walls in line with insulation layers.*

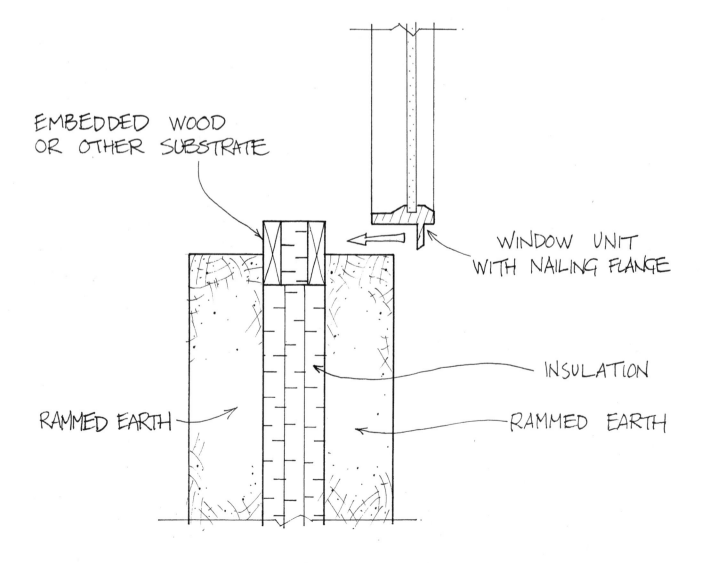

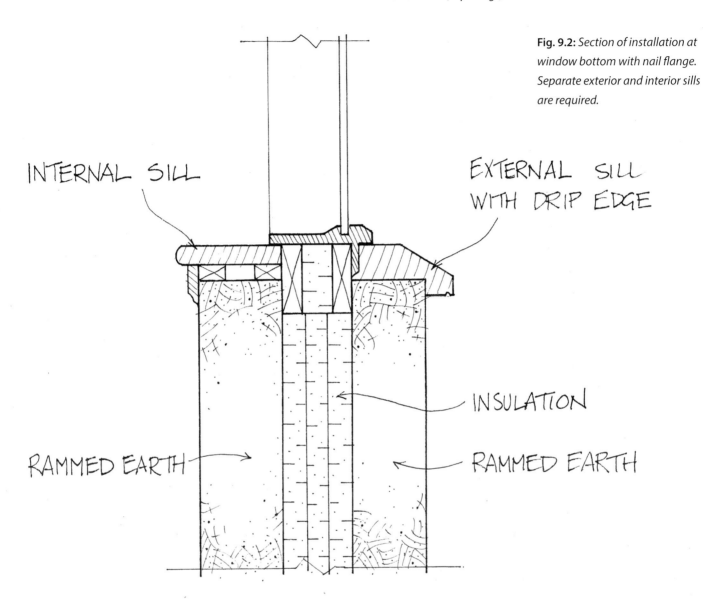

Fig. 9.2: *Section of installation at window bottom with nail flange. Separate exterior and interior sills are required.*

Fig. 9.3: *External sill made of pre-cast concrete. Note metal jamb extensions on sides of opening.* Photo credit: Dalila Seckar

Fig. 9.4: *Metal flashing installed at bottom of exterior side of window opening.* Photo credit: Dalila Seckar

Many modern window frame designs incorporate metal straps as the structural mounting mechanism. This allows some flexibility in situating the glazing within the wall assembly, either closer to one side or the other. Many builders create a buck around the opening in this case, usually with plywood. Figure 9.5 is a photo of the interior side of this style of installation.

At the risk of seeming obsessed with Euclidean geometries, window installation is another situation where achieving square, plumb, and level surfaces with your rammed earth walls is invaluable. The window units you install will be built (or should be) to a very fine tolerance in terms of square and straightness. Attempting to install them into non-square, non-plumb openings with off-level bases can be very frustrating and leads to excessive use of shims and fillers like spray foam to get an acceptable air seal.

The most minimal strategy that I have seen used for window mounting is to use nailer material at set intervals embedded into the sides and top of an opening, located to match the strap locations on the relevant window unit. Figure 9.6 is a photo of this strategy. A major goal of this detail is to minimize thermal bridging and keep the insulation layer as continuous as possible.

Fig. 9.5: *Plywood buck with steel strap installation system. Note shims for location within the opening and foam for air sealing.*

PHOTO CREDIT: DALILA SECKAR

Fig. 9.6: *Embedded nailing strips in window opening. Internal insulation in this project is mineral wool. Discoloration of the insulation material in the lower portion is due to exposure to precipitation.* PHOTO CREDIT: DALILA SECKAR

This installation design also incorporates a smaller exterior side of the opening in comparison with the interior. This is done with air sealing in mind, and the formwork to create this opening needs to be designed with the specific window unit in mind. I encourage builders to come up with better window installation details — I don't know of the "perfect" installation — we can always improve. Another idea to take into consideration is the incorporation of storm windows. In my climate, there are four to six months a year when I don't want to open my windows at all. Adding an external set of glazing for winter used to be the norm in our houses. There's no good reason that storm windows can't be incorporated into new builds, but your wall opening design needs to take them into account.

Depending on the height of the wall and the exposure of any given window, it may be necessary to incorporate flashing above the opening. Figure 9.7 is a photo of a soffit and fascia option at a corner and above two windows in a single-story rammed earth building.

For cement-stabilized rammed earth, it is important to protect the walls from rain during construction to avoid streaking. Some builders recommend cleaning the walls with clean water using a pressure-washer one day after stripping the forms. It is possible to use a wire brush to remove efflorescence; this can be done at any time — but it does change the texture of the surface. If you only brush a small area, that portion will look different under certain light conditions.

There are design and construction challenges at the top of a rammed earth wall. It is difficult to ram a consistent, even horizontal surface — especially to the degree that would be expected for a conventional sill plate installation. Setting the top of wall level is important for roof installation. Shimming individual rafters or trusses after the fact is not ideal. The rammed earth walls are wide enough that a square area for the roof members should occur within their width — but the walls may not be exactly plumb. Level can be controlled or fixed even if square isn't perfect, but depending on what kind of roof finish is being used, there may not be a lot of tolerance.

Air sealing can become an issue at the top of walls, although many builders continue the internal insulation layer up past the top of the rammed earth and connect it directly to the roof or attic insulation. This is highly recommended, as it serves to eliminate thermal bridging at the roof-to-wall transition. It also gives a good opportunity to maintain the air barrier as well.

Generally speaking, rammed earth is not a great surface to build directly on top of. Grout is an option for a cap, or a cast concrete bond beam. The top of the last lift is often not as strong as the rest of the wall due to different curing conditions and the fact that it is not confined from above. Some builders favor creating their own higher-cement, wetter mix using the same materials as the rest of the wall. This mix can be placed and troweled smooth and level in the same manner as the base for the VDB shown in Chapter 7. It is a good idea to test this mix ahead of time for color, texture, and strength.

Fig. 9.7: *Custom metal work at soffit and fascia. Note white staining (efflorescence) in rammed earth; this may be caused by incomplete mixing of cementitious material leaving concentrated amounts of binder which later are exposed to precipitation and begin to hydrate after the rest of the wall has set.*
Photo credit: Dalila Seckar

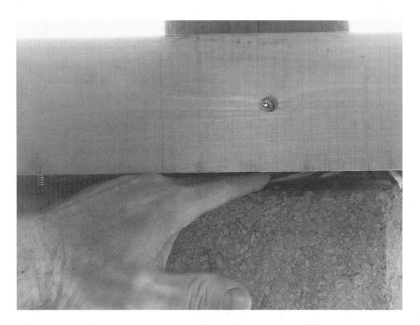

Fig. 9.8: *Photo showing gap between top of interior rammed earth wythe and timber.*

Fig. 9.9: *Interior wood window finish.* Photo credit: Dalila Seckar

If a cementitious bond beam is implemented, it should be installed the same day as the last rammed lift to avoid mix water from the wet mix running down the formwork and staining your wall.

In terms of finish, what about the transition from the top of the wall to the ceiling — do you want to see the bond beam? Do you prefer trim, such as a crown molding? Or perhaps a reveal? Some builders have had success putting a reveal at both the top and bottom of walls to create a straight, even shadow line at the transition from floor to wall and wall to ceiling. A reveal can be created using strips of formply on the inside of the formwork, installed in a similar fashion to the VDB base locators shown in Chapter 7. This technique can also be used to install a screed strip on the inside of the form to the correct level at the top of the wall if putting a cast cap above the rammed earth.

As mentioned earlier, thermal bridging can be avoided by allowing insulation to continue up between the interior and exterior rammed earth wythes. Wood sills made of dimensional lumber can be set and under-filled with grout or other mortar-like material if you want to minimize the depth of any non-rammed earth material at the top of the wall. Threaded rod or j-bolt anchors can be set in upper lifts and used to tie down the sill material. A timber or engineered wood bond beam can bridge both wythes if necessary for structural requirements or to provide a wider bearing surface. Figure 9.8 shows the space between the bottom of a timber sill and the top of a rammed earth interior wythe.

Joints like this need special attention for air sealing, as the wood will expand and contract differently than the rammed earth in response to changing temperature and moisture conditions.

Indoor finish options don't need to incorporate the weather barrier and drainage features that the exterior side does. Figure 9.9 is a photo of the same installation style as Figure 9.5, shown with finish trim completed.

Doors need a firm support at the base. Foam glass is a great option here, but wood also works. An important consideration is the level of thermal performance required. A relatively simple detail is to modify the design shown in Figures 9.1 and 9.2 by having the embedded nailing material level with the opening. This provides an anchor point on either side of the insulation and maintains a thermal break between the rammed earth wythes.

A note about air sealing: Ideally, it is happening at the plane of the insulation, either on the interior or exterior side, but consistently throughout the build. There are tape products available that preferentially adhere to rammed earth, stone, concrete, wood, plastic, and other substrates. Many of these tapes remain flexible and can provide a long-term seal between different materials that expand and contract differentially under varying temperature and moisture conditions.

As mentioned earlier, another consequence of not keeping your rammed earth wall sections aligned is the increased time it takes to do finish work. If you are paying someone to do this work, this means added monetary cost. Most construction materials are manufactured with straight edges, and gypsum wall board, or drywall, is an excellent example. In order to place a piece of drywall against a rammed earth wall and keep a straight line, it may be necessary to incorporate trim, as shown in Figure 9.10.

Embedded Utilities: Chases and Conduit

Electrical boxes and conduit are important embedded elements in rammed earth — unless you are going to go with surface-mounted wiring, which may be an option for your project. If you are embedding the electricals, it is a good idea to stick with single and double boxes for switches and plugs rather than going up to triple and larger ganged-together boxes. Sealed electrical boxes designed for use with masonry are the best option. The heavier metal boxes do not require reinforcement, while lighter ones do. Because it is easy to shatter plastic electrical boxes during tamping, metal is recommended.

It may seem obvious, again, but it is important to know the height of your desired outlets,

Fig. 9.10: *J-trim vertically installed up rammed earth wall to accept edge of drywall. This project also incorporated a dropped ceiling.*
PHOTO CREDIT: DALILA SECKAR

switch, and light locations. For a light switch near a door, it's not a big deal if all of the individual light switches are +/-1″, because they aren't directly beside each other and there isn't a long horizontal line near them for visual reference. But plugs and switches directly above a kitchen or bathroom counter should be at the same level to a very fine tolerance, or it will be obvious to the eye. Kitchen countertops and cabinets are, thus, critical level lines.

Electrical boxes and conduit are installed in the formwork before tamping, as shown in Figure 9.11.

The style of installation shown in Figure 9.11 includes a beveled inset. Figure 9.12 is a photo of a finished light switch installation done using this technique.

Another technique is to attach the box directly to the inside of the formwork to achieve a flush mount. In this system, use the threaded holes in the box to hold the box against the inside of the formwork. You will need to use longer machine thread screws than are included with the box to make this work. For setting the boxes onto the formwork, first create a template that allows for holes spaced at the same dimensions as the electrical box threaded holes. Drill through the formwork at the desired locations. All of these holes can be drilled ahead of time — this is a great task for a detail-oriented member

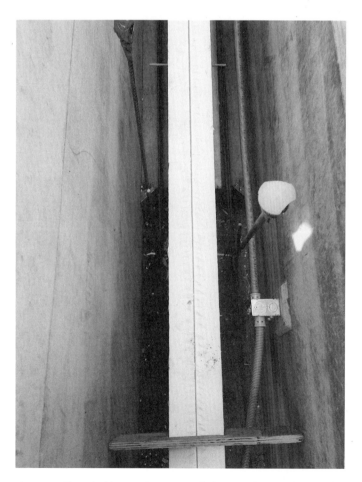

Fig. 9.11: *Electrical box with inset inside formwork. Note the use of both flexible (bottom of photo) and solid conduit (top of photo).* Photo credit: Graham Cavalier

Fig. 9.12: *Finished light switch installation in end of interior wall with beveled inset.* Photo credit: Dalila Seckar

of the crew to work on in between other work flow tasks. Then, when preparing for tamping and installing the boxes and conduit, have one person in the formwork holding the boxes against the formply while another person on the outside puts in the screws. The locations of each box should be marked and all screws removed before stripping the forms, just as with VDBs. It is best to use hex nut or Allen head screws for this, as it is too easy to round off regular screws. Hex head and Allen head screws are higher grade and will resist snapping if the tamper accidentally hits the box. If you are careful, you may even be able to reuse these screws multiple times. Figure 9.13 shows a flush-mount electrical box after the formply has been removed.

When tamping in the lift containing the box, it is good to deliver loose material flush to the top of box. This allows you to work around the box more easily by hand — reach under and make sure there is no loose, large gravel against the forms. Use a small wood tamper to compact the mix close to the box, and use your hands. Get in there and be diligent. This lift is tamped to an elevation below the top of the box. The next lift should be a bit deeper in order to protect the top of the box from damage during compaction.

Figure 9.14 is a photo of an embedded electrical box installation that didn't have all of the material beneath the box properly compacted.

As shown in Figure 9.11, conduit can be solid or flexible, but it is more difficult to fish wire through the flexible conduit than the solid product. It is better to take the time and plan out the entire installation using solid conduit to save time and energy during the electrical rough-in. Planning to minimize electrical runs will help minimize the amount and length of conduit you have to work around inside the formwork. Vertically installed conduit is just like another piece of vertical rebar to a worker inside the forms.

Fig. 9.13: *Flush-mount electrical box in rammed earth wall.* Photo credit: Dalila Seckar

Fig. 9.14: *Triple electrical box with inconsistent compaction at bottom. Note the grout filled in after tamping by owner.*

Fig. 9.15: *Fireplace installed in rammed earth chimney surround.*
Photo credit: Dalila Seckar

Walers and strong-backs can easily get in the way of electrical box installations — plan your electrical layout knowing the horizontal and vertical spacing of these elements of your formwork. Similarly, try to line up through-tie placement with wall or floor locations. You can use the intersection between an interior wall perpendicular to the rammed earth to hide conduit chases, or other wall penetrations.

Plumbing should be avoided in exterior walls, but it is possible to embed pipes in the interior wythe. When creating plumbing chases, it is good to use a larger-diameter pipe to create a conduit, and then run pipes inside that conduit after ramming is finished. Most residential codes require plumbing chases to be placed within the noncombustible material, not in the insulation. Vertical chases can be used to add flexibility and long-term access points. Always follow applicable codes when embedding anything in the rammed earth. There may be minimum coverage depths that are required between the edges of the wall and the conduit, pipe, or wire in question.

There are other penetrations that may be necessary in rammed earth walls, such as hose bibs, ventilation exhaust and makeup air, potable water in, sewer out, electrical feed, communications, and combustible appliance air intake and exhaust, among others. All penetrations should be considered ahead of time and incorporated into the formwork planning and design. Core drilling is possible, but not easy — especially in larger diameters. PVC pipe, ABS, and HDPE products are all available to create the chases or conduits necessary. These types of materials tend to work as permanently embedded chase walls. Steel or sheet metal ductwork with internal reinforcement can also be used. After ramming, the metal can be removed or it can remain in place — it is more of a thermal bridge than plastic or wood embedments. As shown in Chapter 7, other shapes of volume displacement formwork can be created if necessary. As a rule of thumb, multiple smaller-diameter penetrations are easier to ram around than single larger ones.

Surface Treatments and Sealers

Interior rammed earth walls add thermal mass and can be beautiful features in a building. Figures 9.15 and 9.16 are photos of a stabilized rammed earth chimney surround with an inset fireplace unit.

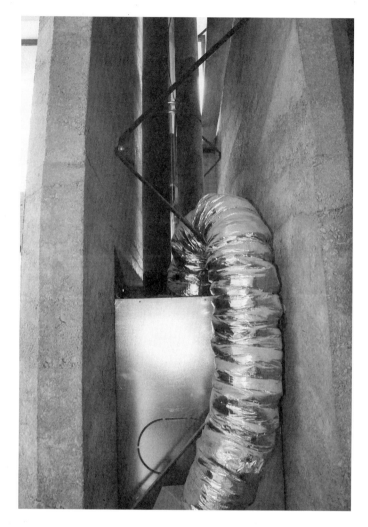

Fig. 9.16 (left): *Side view between rammed earth chimney surround sections. Note steel connections between rammed earth elements and insulated steel chimney pipe, and the combustion air intake.*

Fig. 9.17 (right): *Geoid inset in rammed earth; this example is about 65 mm (2 ½") across.*

Decorative elements can be added to wall surfaces with inclusions or patterned formwork. Figure 9.17 is a photo of a fossil set directly against the formply face and tamped into place. A similar technique can be used to place colored aggregate or other decorative elements against the formply at the top of a finished lift.

Patterned formwork can be used to leave an imprint or carve a pattern into the surface of the formply that the RE then fills — creating an extension of the surface. Relatively easy-to-use CNC technology (*computer numerical control* — used with the better-known CAD programs) can make very accurate patterns in wood surfaces with computer controlled router heads and shapers. Figures 9.18 and 9.19 are photos of a pattern cut into formply, and the resulting image created on the rammed earth wall.

Properly built, internally sealed stabilized rammed earth walls are about as close to maintenance free as anything can reasonably claim to be. There will be some fading over time, and some erosion, staining, and/or efflorescence wherever there is long-term water exposure, similar to masonry walls. So, there is still a need to control drainage around the building, keeping water away from the walls with rainwater leaders, eavestroughs (gutters), and properly sloped landscaping around the perimeter.

Fig. 9.18: *Manually aligning formply joint within void pattern cut into the surface of formply.*
PHOTO CREDIT: CLIFTON SCHOOLEY

Fig. 9.19: *Finished pattern on exposed rammed earth wall.*
PHOTO CREDIT: CLIFTON SCHOOLEY

Rammed earth is a very stable, noncompostable material — you can choose to apply glossy, thick sealers, and it will not harm the basic structure or cause rot. However, if you go that route, it isn't easy to go back. Areas like kitchen or bathroom backsplashes can be treated with marine-grade sealers. Maintenance will be at the interval required by the manufacturer.

External sealers can also be added later and periodically reapplied on either interior or exterior walls. Surface-applied sealers can photodegrade, and some aggressive sealers may yellow or crack over time. I recommend doing research into the products you are considering, and consulting the Living Building Challenge's Red List for nontoxic options.

Many builders have had success with silica emulsion sealers, such as silane or siloxane.

These products are similar in chemical makeup to the internal sealer that I recommended in the mix design portion of the book. Some external sealers are designed to be penetrating and some are designed to be a coating that stays on the surface of the wall. Externally applied sealers will lower the wall's ability to accept and release moisture in the form of humidity, but this may be acceptable for a particular wall area.

It may be that you want the rammed earth to be as low permeability as possible, for instance if you are forming the side of a garden pond. In that case, using internal and external sealers is advised.

Repairs and Renovations

It is possible to repair rammed earth if it was poorly tamped or improperly mixed in the first place. First, remove as much loose material as is safely possible using a wire brush or a similar tool. Then, using a mortar or plaster mix, trowel on or point in wet material and allow it to cure in place. Figure 9.20 is a photo of patching material being applied to a high void area on the exterior side of a stabilized rammed earth wall.

Additions or renovations to rammed earth walls are definitely possible. Techniques used for concrete or masonry wall construction should translate directly to rammed earth, either raw or stabilized. I caution builders to pay attention to thermal bridging, drainage conditions, and differential settlements between old and new sections of buildings when designing and implementing any kind of addition to an existing building.

Fig. 9.20: *Poorly compacted material at the base of overly thick rammed earth lifts is repaired after the forms are removed using a high-strength mortar mix applied by trowel.*
PHOTO CREDIT: DAVID NEUFELD

Relevant Research

Gomes, Maria Idalia and Paulina Faria. "Repair mortars for rammed earth constructions." Proceedings of the International Conference on Durability of Building Materials and Components, Portugal, April 12–15, 2011.

Chapter 10

Building Code Developments

I HAVE A LOVE/HATE RELATIONSHIP with regulations like building codes. On the one hand, they are evolving, living documents that develop in response to public and industry stakeholder input. In Canada, any legal resident is considered a public stakeholder and has access to the committees that adjudicate the national model building code. They also have access to their local municipal council and provincial or territorial committees, giving three points of input. As a registered professional engineer, I can apply to sit on technical committees that write and revise the minutia of the main document and its referenced standards. This does all imply a significant investment of personal time and energy, but it is still possible to participate in the code development process as an individual.

On the other hand, building codes tend to be reactionary, backward-looking documents that are chock-full of bandage-style attempts to fix both human-caused and natural disasters. The language of codes is usually unclear and subject to interpretation, and it is often enforced by persons who seem to believe that no range of interpretation is possible.

My natural building colleague David Eisenberg, head of the Development Center for Appropriate Technology in Tucson, Arizona, pointed out to me that documents like building codes (at least in North America) are anything but designed. This struck me as a good explanation of why the type of person who appreciates good design might not agree with the structure and enforcement of building codes. However, it could be argued that the current trend toward increasing energy-efficiency targets is a more proactive model for code development. Perhaps we'll see more proactive regulations as increased attention to embodied energy and carbon accounting give us metrics to direct us on our way to meeting climate change commitments. As mentioned earlier, we can all get involved — if we choose to be active stakeholders in the development and evolution of our building codes.

That said, the core purposes of codes — to promote safety, health, energy efficiency, and regional appropriateness — are all things I can firmly get behind. Ultimately, builders and code officials have the same goals (or should have), and I have found that a cooperative and collaborative attitude is much healthier and productive than an *act first and ask for forgiveness* approach.

I find it useful to keep in mind how resistant the construction industry as a whole tends to be about accepting new technology, at least until they've seen someone else use it successfully first. By now, builders and regulators in North America are so used to conventional stick frame residential construction using dimensional lumber that failures in these systems are automatically blamed on the builder, not the material. We all know that you can build successfully using dimensional lumber, so if something has gone wrong, it must mean that the builder is at fault rather than the material.

Conversely, when introducing new materials and technologies, we need to remember that if there are failures in the earliest buildings in a given region, those failures will be blamed as much or more on the material itself rather than on the builder.

Canadian and American Building Codes

I've written a number of conference papers about building rammed earth houses within the Canadian building code context, in particular within the province of Ontario. Canada has a national model building code that is adopted by each province and territory after applying amendments, additions, and omissions as they each see fit.

The first National Building Code of Canada (NBCC) to adopt an objective-based format was issued in 2005. The Canadian Commission on Building and Fire Codes attempts to re-issue an updated version of the major codes (Building, Fire, Plumbing, and Electrical) every five years. The current national model building code is the 2015 edition.

The move to an objective-based code did not eliminate the provision of prescriptive solutions for a given building assembly; what it involved, instead, was the addition of *alternative regulatory paths to acceptable solutions*. By defining the goals of the code via cross-referenced objective and functional statements, the objective-based format attempts to give designers and code officials methods to develop and evaluate potential designs for conformance apart from a "cookbook" approach. Specifically, an alternative solutions proposal protocol was first introduced into the 2005 NBCC. However, differences in the way that each province and territory adopts the model code into their legislation, compounded with differences in the way any given municipality enforces their regional code and/or modifies it via local by-laws, leave designers and project proponents with a range of conditions to deal with when applying for a permit to build.

Both before and after the advent of the objective-based code model, a key element to winning the building official's approval directly or via a research program is the establishment of material qualities that can be measured and shown to be consistent with the design methodology adopted by the engineer or architect. This is a primary challenge for the designer: choosing an accepted design methodology developed for a similar, yet different, material and then developing a test method to prove that the different material behaves sufficiently like the accepted one to justify the analysis and final design.

The Canadian Construction Materials Centre (CCMC) is responsible for evaluation and national certification of innovative building materials, products, and systems. Conventionally, a material or product attains CCMC certification in order to be widely accepted by designers, regulators, and builders. All CCMC certifications are referenced in the NBCC by default, allowing relatively easy specification and acceptance. At some point in the future, the material or product may be cited directly in the body of the building code itself.

Polystyrene Insulated Concrete Forms (ICFs) are an example of a product/system that has gone from CCMC evaluation to outright specification in the national code within the past 20 years. This may be frustrating to natural builders who have toiled away for even longer without the resources provided by a large manufacturing association like the Portland Cement Association. However, unless you are very cynical, you have to admit that industry stakeholders do have a place at the building code development table, and that these organizations are simply doing a very good job at putting their products into the best place possible to be adopted widely by the building industry.

The development and acceptance of the Strawbale Construction guidelines in the International Residential Code Appendix in 2015 is testament to a much smaller, less well-funded group of people successfully getting their material of choice into a national building code. A big shout out to Martin Hammer, and (again) David Eisenberg, as well as many contributors from the CASBA (California Straw Building Association) community. Similarly, the development and publication of the ASTM E2392 Standard Guide for Design of Earthen Wall Building Systems represents a tremendous value in terms of monetary investment and a huge effort from a relatively small group of dedicated people, including, notably, Bruce King.

That said, the variability in aggregate content, mix recipe, and methods inherent in working with natural or pre-industrial materials and techniques such as rammed earth effectively precludes its evaluation by a body like the CCMC. It should also be noted that the CCMC evaluation process is lengthy and expensive; to date, there have not been any proponents of earthen construction in Canada willing to attempt it.

This leads to each project being evaluated on its own merits, which raises another challenge. Before the issuing of a permit, inspection criteria must be determined along with a quality control and materials testing program to be carried out during construction. A requirement for any project that varies from common construction techniques or materials is a written commitment to general review(s) by the registered professional(s) responsible for the design. This tends to be a basic form establishing the party or parties responsible for inspections and site reviews, but it does not include a great amount of detail and often needs to be accompanied by a document clearly stating the agreed-upon schedule, notable milestones, and "substantial completion" criteria.

Alternative Solutions Proposals

There are several different methods that authorities having jurisdiction may require a proponent to use in order to incorporate an alternative solution proposal into a permit application package. Some jurisdictions provide a form that gives a format to follow, some have a checklist, and some take a case-by-case approach.

What unites all alternative solutions proposals here in Canada is the need to refer to the relevant objective and functional statements from the provincial building code — along with an argument for why the proposed solution is equivalent or superior to an accepted solution. Documentation may also be required. This can include, but is not limited to: test results, manufacturer's technical specifications, academic papers, precedents, and/or professional analysis.

A volunteer group of stakeholders in British Columbia formed the Alternative Solutions Resource Institute a number of years back to support builders wanting to work with non-conventional materials, in particular straw bale. They published a document with guidance on preparing an alternative solutions proposal, and if you can find a copy, I highly recommend reading through it. You can find a review of it at: http://thelaststraw.org/publication-review-the-straw-bale-alternative-solutions-resource-by-asri/.

As with building design, I do not believe in a "canned" approach to regulatory compliance. Early attempts at alternative solutions proposals that I was involved with went for a comprehensive approach — trying to answer every possible line of the building code that might apply to the design. In some cases this was effective, but it sometimes raised more questions than necessary.

It may be a better strategy to address specific elements of the overall design. For instance, if you are planning on using the rammed earth as your air/vapor barrier and do not want to install a conventional vapor barrier in your wall system, you will need to address this.

An alternative solutions proposal for a wall system without a conforming vapor barrier layer that was presented to the 2012 Ontario Building Code is given in Appendix B. Consider using this as a general guide. Similar proposals could be made for alternative insulation materials, for defending the location of the air barrier or insulation within the wall assembly, or for the air barrier performance of rammed earth. It is best to work together with your building official to make your submission as concise and specifically directed as possible. Officials may have had experience with other proposals involving products that have not been tested in their jurisdiction. For instance, many high-performance window and door products from Europe have not been tested to Canadian or US standards. That doesn't mean they won't work in our climate, but a formal argument for their use in your project may be required — and you may not be the first person in your area to go down that path.

Other Notable Building Codes, Guidelines, and Standards

This list is by no means exhaustive; more references are given in the bibliography section. That said, these documents are the ones that I believe are most relevant to a builder considering earthen construction in North America.

Australia: Bulletin 5 was Australia's national reference document for adobe, compressed earth block, and rammed earth construction for decades. G.F. Middleton, an assistant technical officer at the Commonwealth Experimental Building Station in New South Wales, was the principal author of Bulletin 5 in 1952. The handbook was updated in 1976, 1981, and finally in 1987.

In the 1990s, a cooperative project between the Earth Building Association of Australia and their counterparts, the Earth Building Association of New Zealand, was undertaken to update the standard. The two groups did not arrive at consensus, and New Zealand published their own standard in 1998.

The Earth Building Association of Australia published a draft document of proposed design guidelines for adobe and rammed earth in 2001.

In 2002, The Australian Earth Building Handbook was published by Standards Australia. Also known as HB 195, the content of this document was prepared jointly by the standards organization committee BD-083 and Dr. Pete Walker, a prominent earthen construction researcher and author from the University of Bath, UK.

HB 195 (2002) is an advisory handbook; it is not a legal Australian standard. It is available for purchase from SAI Global. An update is in the works.

New Zealand: After failing to find consensus with their colleagues from Australia, the Earth Building Association of New Zealand was able to have a set of three standards published by Standards New Zealand in 1998. They are NZS 4297: 1998 Engineering Design of Earth Buildings; NZS 4298: 1998 Materials and Workmanship for Earth Buildings; and NZS 4299: 1998 Earth Buildings Not Requiring Specific Design.

This set of standards is still available from Standards New Zealand, and in my opinion these standards are one of the most accessible and sensible set of code documents for earthen construction available today.

Germany: There is a rich history of earthen construction in Germany, with guidelines formally published in 1947 and 1956 in West Germany. These were withdrawn in 1970 and nothing replaced them until 1999 with the publication of the Lehmbau Regeln. This document does not give a code per se, but it has been referenced by several regional governments in Germany. The guideline was updated in 2002 and 2009.

United States (State of New Mexico): New Mexico first published their own earthen building code in 1991. At that time, the Universal Building Code was still the parent document for many state building codes. Now the International Building Code and International Residential Code are the national reference codes that are adopted by individual states.

The New Mexico Standard for Rammed Earth Construction is a concise document with empirical design guidelines for small residential buildings. In 2006, the document was updated as the New Mexico Earthen Building Materials Code.

United States (Pima County, State of Arizona): This jurisdiction publishes amendments to the International Building Code for both straw bale and earthen construction. Tucson is in Pima County.

American Society for Testing and Materials (ASTM International): ASTM E2392/E2392M Standard Guide for Design of Earthen Wall Building Systems, was first published in 2005. The standard provides guidelines for earthen construction using several techniques, including rammed earth. It references a number of other relevant standards from the US, as well as the New Zealand standards.

For many builders in North America, being able to refer to a standard published by a nationally respected organization like ASTM can be a game changer for having a design granted a building permit. The standard contains useful definitions, best practices, empirical design guidelines, and a summary of testing methods for earthen materials. ASTM E2392 was updated in 2010.

Many other countries have strong earthen building communities as well. CRAterre in France is an excellent organization, truly an international center for earthen construction. They collaborate with the Auroville Earth Institute in India under the auspices of UNESCO. Both of these organizations offer education and research opportunities, and they publish guides, academic research, and newsletters on a regular basis.

The Indian Institute of Science's Center for Sustainable Technologies has carried out a considerable amount of research into earthen construction, and it continues to move the state of the art forward. Sri Lanka has published a set of standards for building with compressed earth block, the SLS 1382, parts 1 through 3.

In Africa, Zimbabwe has a national standard for rammed earth structures, first published in 2001. It is based largely on Julian Keable's *Code of Practice for Rammed Earth Structures* from 1996.

Tanzania, Mozambique, Morocco, Tunisia, Kenya, and Ivory Coast are other African nations that have had earthen construction standards at some point in time. Many of these documents have gone out of print and are difficult to find today.

Appendix A: Sample Engineering Specification

Preface

SPECIFICATIONS ARE one of many contract documents common in commercial construction. They are important for defining scope of work, material tolerances, performance definitions, and just about anything else that project team members may need to reach agreement on. Just about any material or product called out in a design may be required to have a specification written up for that particular project. Some specifications may be more generic, such as for a given type of roofing material, or metal flashing.

In some cases, a separate document may not be necessary. For example, dimension lumber can simply be called out in a set of notes directly on construction drawings with species and grade required. Anyone requiring further clarity can then look up the specification published by the grading authority for that wood species and the lumber milled from it.

Specifications are rarely required for permit application purposes in residential construction, but it is within the rights of a building official to ask for technical information on any material not explicitly defined in their building code or local by-laws.

The purpose of our specification is to identify the key quality points of the stabilized soil mix, formwork, reinforcing, placement procedures, and all other relevant materials and accessories required to achieve the project targets with respect to the rammed earth.

Stabilized Rammed Earth Engineering Manufacturing Specification

Part 1 General

1.1 Related Documents

A. Permit Drawings/Package

1.2 Submittals

A. Submit laboratory compressive strength test reports for pre-construction mix qualification to the structural engineer and geotechnical engineer for review prior to construction, (refer to 4.1 A).

B. Structural Drawings: Indicate expected minimum compressive strength performance, hydroscopic and high binder content locations, reinforcing schedule, waterproofing details, affected related work, expansion/contraction joint location.

Structural drawings to be signed and sealed by an engineer licensed to practice in the local jurisdiction.

In our example, the specification is being provided by the design team as part of the tender package. Bidders seeking to win the contract to build the rammed earth portion of the project will use the specification and related documents as the guideline for their submittal package.

The specification is to be read in conjunction with the permit drawings and any other documents the design team provides, ranging from city planning documents to federal/regional environmental regulations.

Submittals may be documents or physical specimens, as with color or texture samples.

C. Color/Texture samples to be prepared and reviewed in conjunction with project design team. Coordinate sample preparation with compressive strength test schedule.

Part 2 Materials

2.1 Form Materials

A. Forming materials that create the exposed finished surface to be plywood, metal, or other acceptable panel-type materials to provide continuous, smooth, straight walls. Finish in largest practical sizes to minimize number of joints.

B. Coordinate with project design team on desired finish and method for any given section of the project.

C. Form release agent: provide commercial formulation form release agent with a maximum of 350 g/L volatile organic compounds (VOCs) that will not bond with, stain, or adversely affect stabilized earth surfaces and will not impair subsequent treatments of surfaces.

D. Bracing materials to be sufficient to bear and react vertical, lateral, static and dynamic compaction forces without opening unacceptable gaps or shifting from level and plumb.

> The project that our example specification is from has a variety of different finish textures required by the lead designer, a landscape architect. Because of this requirement, a variety of forming materials are allowed, and because there isn't a "tried and true" method for achieving the designer's desired range of textures, there is a certain degree of ambiguity in specification.
>
> For most residential projects using a paper-faced formply, no release agent will be necessary. It's included in the example because the project in question has a wide variety of finish textures in the design.

2.2 Reinforcing Materials

A. Material size, type, and schedule to be specified within structural drawings.

B. Reinforcing connected to other structural or foundation components to be coordinated with concrete or mechanically stabilized earth contractors.

2.3 Stabilized Soil Materials

A. Portland Cement: CAN/CSA A3001

B. Lime: CAN/CSA A179-14

C. Other pozzolans: Calcined clays or naturally occurring pozzolans to be specified on a case by case basis

D. Color: Oxides and mix proportions as determined by color samples

E. Soil: As determined by geotechnical firm specifications. Typical specification includes quarry and product name, maximum aggregate size, maximum % passing #200 sieve, plastic limit, liquid limit, pH, and plasticity index. Frequency of material property testing based on geotechnical engineering firm schedule. Materials to be free of organic materials and other contaminants.

F. Silicate emulsion to reduce permeability and efflorescence — to manufacturers specification

G. Water: clear, free from visible sediment and color, preferably potable.

> The example project does not include any indoor or conditioned spaces, so the specification does not include anything for insulation or other building envelope characteristics such as desired vapor permeability or maximum air leakage rates. If these are important project goals for your build, and you are using a specification like this to help choose contractors, this is the place in your specification to include desired insulation options.

2.4 Nailing Strips

> This could also be titled "Permanent Embedments", as it can cover just about anything embedded in the rammed earth to provide an attachment opportunity.

A. Permanent embedded nailing strips to be non-organic material, plastic grade lumber, steel extrusions or other acceptable materials to provide a rigid and durable connection substrate.

B. Replaceable nailing strips can be of organic material provided the method of attachment allows for replacement without damage to the cured walls.

C. Anchorage for permanent structures such as guard rails to be coordinated with supplier and installer.

Part 3 Executions

3.1 Mixing

> As described in the text, you may have relatively wet aggregate from your supplier, and if so, you should only start with 5 or 6% (+/- 1%) of the mix weight to begin with and work your way up from there as necessary.

A. Mixing must be performed on a clean, dry, on-site surface kept continuously clean of extraneous debris and partially cured cement-stabilized earth. Material shall be placed and rammed within 1.0 hour of wetting.

B. In the absence of mix specific optimum moisture content test results, mix water to 10% (+/- 1%) of the dry mix weight (this is the amount of water that when added and mixed, will give a material consistency that can be balled up in the hand, dropped from waist height, and will break into several coherent clods neither shattering nor splatting — the traditional "ball drop" test)

3.2 Pre-soil placement

A. Perform and verify all existing conditions before starting placement; all dimensions and locations required on drawings, anchors, seats, plates, reinforcement, joint devices, and other items to be cast into rammed earth are accurately placed, positioned securely, and will not impede rammed earth placement. Locations of all openings and embedments required for other structural, architectural and electrical work are in place and waterproofing of foundation wall has been applied. Ensure formwork is clean, and that any chips and debris have been cleared from footing or previous lift.

3.3 Soil Placement

A. Install reinforcing according to structural drawings. Install individual lift for full length of wall or to the maximum extent covered by a single batch to depth of 100 mm – 200 mm and compress with pneumatic tampers. Ensure reinforcement, inserts, embedded parts, and expansion and contraction devices are not disturbed during rammed earth placement.

B. Coordinate mix placement with appropriate section of exposed face as determined by project team in color/texture approval process.

3.4 Curing and Protection

A. Immediately after placement, protect rammed earth from rain and flowing water, premature drying, excessively hot or cold temperatures, and mechanical damage.

B. During the curing process protect cement stabilized soil from physical damage or reduced strength that could be caused by frost, freezing actions, or low temperatures. Temperatures below 4°C (39°F) can significantly slow hydration rates in uncured stabilized soil. An external heat source or insulation should be employed to ensure uncured stabilized earth temperatures do not fall below 0°C (32°F).

> A residential project may not have a geotechnical engineering firm involved in the design or specification portions of the build. The test methods given in this book may be sufficient to find an acceptable mix and to carry out the project at a high enough quality to satisfy the local building authority. A larger commercial project will definitely have a geotechnical firm on the design team.

3.5 Removing Formwork

> Rammed earth can generally be considered self-supporting as soon as compaction is completed — but that doesn't mean that it is a good idea to strip formwork immediately. The surface of the walls will be very soft and susceptible to damage during the removal of formply, especially at the edges/corners of VDBs or other openings.

A. Formwork not supporting weight of stabilized soil may be removed after cumulative curing at not less than 10°C (50°F) for 12 hours after placing material, provided stabilized soil is sufficiently hard to not be damaged by form removal operations, and provided curing and protection operations are maintained.

B. Coordinate formwork removal and subsequent surface treatment with textures approved by project team.

Part 4 Quality Assurance

4.1 General

> The pre-construction and construction phase testing program listed in this specification is for a commercial project in a jurisdiction not familiar with stabilized rammed earth. A smaller, residential scale project will not likely need this level of testing and quality control — but if you have the time and budget, it's always good to make samples and test various recipes ahead of time.
>
> That said, in some cases, this level of testing and strength confirmation may be necessary in order to obtain a building permit and then an occupancy permit after the project is complete.

A. Quarried products must be from the same pit for the entire scope of the project. Minimum acceptable practice would ensure materials come out of a pit region from which material analysis samples were obtained. Best practices involve the use of a dedicated and set-aside quantity of known material made (screened, mixed, washed, etc.) according to specification.

4.2 Mix Design and Pre-Construction Qualification

A. Engage an engineering agency with sufficient geotechnical capability to design mix using local or readily available product to meet minimum specified compressive strength.

B. Pre-qualify compressive strength of mix before start of construction by a program agreed to by the Prime Consultant.

C. Engage a testing agency capable of performing materials testing in accordance with industry-accepted standards ASTM D1633 00(2007) (Standard Test Methods for Compressive Strengths of Molded Soil Cement Cylinders), or ASTM C39 (Standard Test Method for Compressive Strength of Cylindrical Concrete Specimens). Compression test cylinders comply with ASTM C31 (Standard Practice for Making and Curing Concrete Test Specimens in the Field).

D. Typical pre-construction program may consist of nine samples from one mix. Test three samples each at 14, 28, and 56 days. Strength level of pre-construction mix will be considered satisfactory if the average of the three, 56 day samples equal or exceed the minimum specified compressive strength and no individual sample falls below the minimum specified compressive strength. 14 and 28 day test data will generate curing curves which will be used to relate early age construction strength test results to 56 day strength.

4.3 Construction Compressive Strength Test Program

A. Test program agreed to the Prime Consultant. Typical construction compression test program to consist of one sample from each mix. Early indication that compressive strength of construction mixes will achieve final 56 day specification is determined from 14 day test data falling within a 10% error band fitted to the curing curve. If the 14 day test data fall outside a 10% error band a full review of construction mixing method and material analysis will be carried out to determine any deviations from pre-construction mixing method and earth material specification.

Appendix B: Alternative Solutions Proposal

Sample Alternative Solution Proposal:

Rammed earth finish for walls above grade, forming part of a continuous air barrier, with vapor permeability below 60 ng/(Pa.s.m^2), but not measured in accordance with ASTM E96/E96M.

Description of proposed alternative solution:

Final wall assembly to consist of an exterior stabilized rammed earth wythe of no less than 150 mm in thickness, internal rigid insulation of no less than 150 mm in depth (minimum R24), and an interior stabilized rammed earth wythe of no less than 200 mm in thickness. The air barrier is detailed to occur at the warm side of the insulation layer.

Reason for the request of the alternative solution:

Rammed earth walls, when constructed properly and detailed adequately around openings like doors and windows and at the top and bottom of the wall, provide a continuous air barrier meeting the requirements of OBC 2012 9.25.3.2 (1). The rammed earth does serve as a vapor diffusion retarder, but has not been tested for conformance to the limit required by 9.25.4.2 (1) (i.e. 60 ng/[Pa.s.m^2] when measured in accordance with ASTM E96).

The design proposed does conform to the principles of 9.25.4.3; the wall assembly protects the warm side of the wall and prevents condensation at the design temperature and humidity range. Similar wall assemblies in similar climatic conditions have been observed to maintain safe average annual moisture contents.

The 2012 OBC contains several objective statements regarding sustainability, resource conservation, and environmental stewardship. In particular, objective statements OR and OE are relevant to this proposal:

Objective Statement OR: An objective of this Code is to limit the probability that, as a result of the design or *construction of a building*, a resource will be exposed to an unacceptable risk of depletion or the capacity of the infrastructure supporting the use, treatment, or disposal of the resource will be exposed to an unacceptable risk of being exceeded, and Objective Statement OE: An objective of this Code is to limit the probability that, as a result of the design, *construction* or operation of a *building*, the natural environment will be exposed to an unacceptable risk of degradation.

At this point in time these objectives are not paired with functional statements in the OBC; it is up to practitioners to assert designs, materials, and methodologies that meet these objectives. Use of agricultural by-products, which would otherwise be wasted and potentially add to atmospheric carbon or particulate air pollution (as when they are burned by farmers who cannot process or market them adequately), meets the objective OR. Similarly, use of agricultural by-products in construction conserves habitat otherwise depleted by large scale forestry practices, which meets the objective OE.

Division B provision affected:

9.25.4.2. Vapour Barrier Materials

(1) Vapour barriers shall have a permeance not greater than 60 ng/(Pa.s.m^2), measured in accordance with ASTM E96/E96M, "Water Vapor Transmission of Materials," using the desiccant method (dry cup).

Applicable linked Objective and Functional Statements for 9.25.4.2 (1):

Linked pair A: (F63 — OH1.1, OH1.2)

Functional Statement F63: To limit moisture condensation paired with Objective Statements OH1.1: An objective of this Code is to limit the probability that, as a result of the design or *construction* of a *building*, a person in the *building* will be exposed to an unacceptable risk of illness due to indoor conditions caused by inadequate indoor air quality, and OH1.2: An objective of this Code is to limit the probability that, as a result of the design or *construction* of a *building*, a person in the *building* will be exposed to an unacceptable risk of illness due to indoor conditions caused by inadequate thermal comfort.

Linked pair B: (F63 — OS2.3)

Functional Statement F63: To limit moisture condensation paired with Objective Statement OS2.3: An objective of this Code is to limit the probability that, as a result of the design or *construction* of a *building*, a person in or adjacent to the *building* will be exposed to an unacceptable risk of injury due to structural failure caused by damage to or deterioration of *building* elements.

Both of these objective and functional statement pairs relate to the need for the wall assembly to be constructed in such a way that condensation is avoided and deterioration is resisted. In all cases, these functions and objectives are met by controlling the rate of air leakage through the wall assembly as prescribed in the OBC for building assemblies, and by allowing the wall system to maintain a safe moisture content over the course of an annual weather and occupancy cycle.

List of acceptable solutions in Division B:

9.25.4.2. Vapour Barrier Materials, 9.25.4.3. Installation of Vapour Barriers

The proposed alternative solution meets the intent of 9.25.4.3., but without confirming the explicit permeance limit prescribed in 9.25.4.2.

Information, Documentation, other Former Approvals:

CMHC Technical Series 09–105, Research Highlight "Understanding Vapour Permeance and Condensation in Wall Assemblies"

Definitions

Atterberg limits — The somewhat arbitrary laboratory determined upper and lower limits of moisture content that describe the plasticity and brittleness of a fine grained soil such as clay.

Clay — Soil particles smaller than 0.002 mm (0.000079″).

Formply — Plywood specifically manufactured for use as facing material in concrete formwork. The face of the plywood is either coated with a paper layer or infused with an epoxy to prevent cementitious material from sticking to each sheet. The edges of the formply are treated as well.

Gravel — Soil particles larger than 2 mm (0.079″) in diameter, usually up to 60 mm (2⅜″) diameter.

Lift — A single course or layer of rammed earth.

Limit states design — A structural engineering term used to compare imposed loads to material resistance (aka Load and Resistance Factor Design [LRFD]).

Liquid limit — The upper Atterberg limit — describes the moisture content when a clay begins to flow like a liquid.

Plastic limit — The lower Atterberg limit — describes the moisture content when a clay becomes brittle, or is no longer plastic.

Pozzolanic reaction — The essential reaction in lime stabilization, where the silica and alumina clay minerals undergo dissolution in a high alkaline environment and then recombine with calcium to form complex aluminum and calcium silicates — cement. Pozzolanic binders may be naturally occurring, the by-product of an industrial process such as fly-ash from coal power generation, slag from steel production, or manufactured directly such as calcined clay. The word *pozzolan* is based on the Italian Pozzuoli, a region in Italy with volcanic soils that the Romans discovered and used to make their famous Roman cement.

Sand — Soil particles between 0.06 mm (0.0024″) and 2 mm (0.079″) in diameter.

Sieve — A wire mesh with a known opening size, used to grade soil particles. Sieves are provided in sets from very fine to coarse, with sizes specified by various organizations.

Silt — Soil particles between 0.002 mm (0.000079″) and 0.06 mm (0.0024″) in diameter.

Strong-back — A vertical structural element of a formwork assembly, attached to the outside of the walers, keeping the formwork plumb.

Tamper — A hardened tool for compacting loose earth. Can be manual, pneumatic, or mechanically operated and may be made of wood or metal (aka rammer).

Waler — A horizontal structural element of a formwork assembly, usually attached directly to the forming material that contains the rammed earth.

Wythe — A vertically continuous layer of masonry in a wall.

Resources and Material Suppliers

Aggregate Mix:

Contact local gravel suppliers, check your own site materials

Air Compressor:

XAIR SC70(this model can run 3 tampers @ 20cfm)

Aluminum Formwork:

Aluma Systems Stud Form: aluma.com/us/products/formwork_shoring/wall_formwork/studform

Calcined Clay:

Metapor by Poraver: poraver.com/en/metapor/

Color:

Interstar Ready Mix Colors — pure iron oxide pigments designed to be used specifically in integral coloring of cementitious materials: interstar.ca

Other integral color manufacturers do have product available. I recommend making sure they are certified ASTM C979 for integral coloring.

Fiberglass Rebar:

Tuf-Bar Canada, Erin, Ontario: tuf-bar.com

Geotextiles:

Landscape supply stores; larger suppliers that handle materials for retaining walls, road reinforcement and large scale drainage/civil works.

Mixing:

Telehandler mounted mixing bucket from M3: m3srl.com/eng/products/concrete_mixing_bucket.html

Standalone horizontal axis mixer from EZGrout (MudHog model): ezgmfg.com/masonry-and-construction/grout-concrete-mortar-mixers/

Portland Cement, Hydrated Lime:

Local masonry and concrete supply stores, most building supply stores

Reinforcing Steel:

Local building supply

Tampers:

Henry Air Tools (model 1350-2BF with the steel butt): henrytools.com/rammers-airtools.html

Ingersoll Rand has a similar style tamper available: intlairtool.com/ingersoll-rand-241a1m-floor-sand-rammer/

Chicago Pneumatics, CAPCO, and Atlas Copco are also manufacturer options for tampers.

Waterproofing:

Waterproofing additive: Plasticure by Tech-Dry Building Protection Systems, Melbourne, Australia

Other waterproofing additives are available from Kryton: kryton.com

Bibliography

ASTM International. Standard Guide for Design of Earthen Wall Building Systems, ASTM E2392/E2392M-10. West Conshohocken, PA, 2010.

Baker-Laporte, Paula and Robert Laporte. *EcoNest: Creating Sustainable Sanctuaries of Clay, Straw and Timber.* Gibbs Smith, 2005.

Canadian Commission on Building and Fire Codes. National Building Code of Canada, National Research Council of Canada, Ottawa, 2015.

Canadian Foundation Engineering Manual, 4th edition. Canadian Geotechnical Society, 2006.

Canadian Standards Association. CAN/CSA A23.3–04 — Design of concrete structures. Canadian Standards Association, Mississauga, 2007.

Canadian Standards Association. CAN/CSA S304.1–04 — Design of masonry structures. Canadian Standards Association, Mississauga, 2005.

Ciancio, Daniela and Christopher Beckett, editors. *Rammed Earth Construction: Cutting-Edge Research on Traditional and Modern Rammed Earth,* Proceedings of the 1st International Conference on Rammed Earth Construction, Australia. CRC Press, 2015.

Craig. R.F. *Craig's Soil Mechanics.* Spon Press, 1997.

Crimmel, Sukita Reay and James Thomson. *Earthen Floors: A Modern Approach to an Ancient Practice.* New Society Publishers, 2014.

Drysdale, Robert G. and Ahmad A. Hamid. "Masonry Structures Behaviour and Design." Canada Masonry Design Centre, Ontario, 2005.

Easton, David. *The Rammed Earth House.* Chelsea Green Publishing, 2007.

Hall, Matthew R., Rick Lindsay, and Meror Krayenhoff, editors. *Modern Earth Buildings: Materials, Engineering, Construction and Applications.* Woodhead Publishing, 2012.

Walker, P. and Standards Australia. HB 195 in "The Australian earth building handbook." Standards Australia, Sydney, Australia, 2001.

Heyman, Jacques. *The Stone Skeleton: Structural Engineering of Masonry Architecture.* Cambridge University Press, 1997.

Houben, Hugo and Hubert Guillaud. *Earth Construction: A Comprehensive Guide.* Intermediate Technology Publications, London, 1994.

Hunter, Kaki and Donald Kiffmeyer. *Earthbag Building: The Tools, Tricks and Techniques.* New Society Publishers, 2004.

Jaquin, Paul and Charles Augarde. *Earth Building: History, Science and Conservation.* IHS BRE Press, Berkshire, UK, 2012.

Kachadorian, James. *The Passive Solar House: Using Solar Design to Heat & Cool Your Home.* Chelsea Green Publishing, 1997.

Keable, Julian and Rowland Keable. *Rammed Earth Structures: A Code of Practice.* Practical Action Publishing Ltd, 2011.

Khalili, Nader. *Ceramic Houses and Earth Architecture.* Cal-Earth Press, 2005.

King, Bruce. *Buildings of Earth and Straw: Structural Design for Rammed Earth and Straw-Bale Architecture.* Green Building Press, 1996.

King, Bruce. *Making Better Concrete.* Green Building Press, 2005.

King, Bruce, et al. *The New Carbon Architecture: Building to Cool the Climate.* New Society Publishers, 2017.

Magwood, Chris. *Making Better Buildings: A Comparative Guide to Sustainable Construction for Homeowners and Contractors.* New Society Publishers, 2014.

Maniatidis, Vasilios and Peter Walker. "A Review of Rammed Earth Construction," Natural Building Technology Group, Department of Architecture and Civil Engineering, University of Bath, UK, May 2003.

Mazria, Edward. *The Passive Solar Energy Book.* Rodale Press, 1979.

McHenry, Jr., Paul Graham. *Adobe and Rammed Earth Buildings: Design and Construction.* The University of Arizona Press, 1984.

Middleton, G.F. *Build Your House of Earth.* Angus and Robertson, Sydney, Australia, 1953.

Minke, Gernot. *Building with Earth: Design and Technology of a Sustainable Architecture.* Birkhauser, 2009.

OBC 2012. Ontario Building Code, Ministry of Municipal Affairs and Housing, Markham.

O'Connor, John F. *The Adobe Book.* Ancient City Press, 1973.

Racusin, Jacob Deva and Ace McArleton. *The Natural Building Companion: A Comprehensive Guide to Integrative Design and Construction.* Chelsea Green Publishing, 2012.

Sauer, Marko and Otto Kapfinger, editors. "Martin Rauch: Refined Earth Construction & Design with Rammed Earth." Institute for International Architecture Documentation GmbH & Co, KG. Munich, 2015.

Schroeder, Lisa and Vince Ogletree. *Adobe Homes for All Climates: Simple, Affordable and Earthquake-Resistant Natural Building Techniques.* Chelsea Green Publishing, 2010.

Standards New Zealand. Earth Building Standard NZS4297–1998. Wellington. 1998.

Volhard, Franz. *Light Earth Building: A Handbook for Building with Wood and Earth.* Birkhauser, 2016.

Walker, Peter, Rowland Keable, Joe Martin, and Vasilios Maniatidis. *Rammed Earth: Design and Construction Guidelines.* BRE Bookshop, 2005.

Weisman, Adam and Katy Bryce. *Building with Cob: A Step by Step Guide.* Green Books Ltd., 2007.

Williams-Ellis, Clough, John Eastwick-Field, and Elizabeth Eastwick-Field. *Building in Cob, Pisé, and Stabilized Earth.* Country Life Limited, 1916. Reprinted by Donhead Publishing Ltd., 1999.

Index

Page numbers in *italics* indicate figures and tables.

A

additives, types of, 3
   See also stabilized rammed earth
admixtures. *See* colors; sealers
aesthetics, 91, 105
Africa, standards and codes, 123
aggregate
   blend tests, 32–33
      ingredient measurements, 71
      moisture content, 49
   See also mixes
air compressors, 78–79
air control layer, 13
air sealing, 109, 111
air tampers, 78, 97
alternative solutions proposals, 120, 121–122, 131–132
Alternative Solutions Resource Institute, 121
American Plywood Association (APA), 81, 85
American Society for Testing and Materials (ASTM International), 123
anchor bolts, 67–69
an-isotropic, 43
Arizona, standards and codes, 123
ASTM E2392 Standard Guide for Design of Earthen Wall Building Systems, 63, 121, 123
asymmetrical double-wythe walls, 60, *61*
Atterberg limits, 26–27, 45
Auroville Earth Institute, 123
Australia, standards and codes, 122
The Australian Earth Building Handbook, 122
autoclaved aerated concrete, 60

B

ball drop test, 35–36, *35*, *36*, 48–49, 74
basalt-based rebar, 59
basements, 9
bentonite, 27
binders
   aggregate mix grades and, 29–30
   blend tests, with aggregate, 32–33
   clay characteristics and, 27
   for durability, 69
   ingredient measurements, 71
   types of, 3
bolts, embedded, 67–69, *68*
bond beams, 57, 110
bored samples, 49, 51
building codes
   alternative solutions proposals, 120, 121–122, 131–132
   empirical structural guidelines, 62
   evolution of, 119
   international resources for, 120–121, 122–123
   masonry design codes, 4–5
building procedure. *See* construction
buildings, assembly performance, 9
Bulletin 5, 122

C

calcined clay, 31
California Straw Building Association (CASBA), 121
Canadian Commission on Building and Fire Codes, 120
Canadian Construction Materials Centre (CCMC), 120–121
Canadian Plywood Association (CANPLY), 81–82
Canadian Wood Council, 87
carbon, 5–6, 7
casein, 3
cementitious binders, 3
   See also binders

cement-stabilized rammed earth cost estimates, 101–103 *See also* stabilized rammed earth
Center for Sustainable Technologies, Indian Institute of Science, 123
clay
    binding ability of, 22
    plasticity, 26–27
    structure, *25, 26*
    types of, 24
codes. *See* building codes
coil-rods, 83, 85, 91
coil-ties, 83, 91
colors, 31, 33, 71, 73
compaction test, 41–42, *41*
compressive strength test, 45–48, 49, 51
compressive stress, definition, 65
conduit, 111–113
construction
    cleaning walls, 109
    general techniques, 97–99
    ingredient measurements, 71
    mixing order, 72–73
    See also formwork
construction phase testing, 48–51
control layers, 10
corners, outside, 92, 94–95, *94*
cost estimates, 101–103, *101, 102, 103*
CRAterre, 123
crews/volunteers
    cost estimates, 101, *101,* 102
    experience of, 100
    roles of, 99
cross ties, 85
CSA S304.1 Design of Masonry Structures, 4, 47, 63, 64, 69
curved walls, 58
cut test, 37
cylinders, 45–46, *45, 46,* 49

D

decorative elements, 115
deep form method, 98–99
deflection limit, 87
density, 62
design
    cost estimates, *101,* 102
    drawings, 69–70
    finishes, 57–58, 115
    lintels, 69
    raw rammed earth guidelines, 62–69
    stabilized rammed earth guidelines, 69
"Design/Construction Guide: Concrete Forming" (APA), 85
diaphragms, 67, *67*
door openings, 105
double-wythe walls, 58–60, *59, 61*
drainage, 115
drywall, 111
ductwork, 114
durability, 69

E

Earth Building Association of Australia, 122
Earth Building Association of New Zealand, 122
earthbag building, 30
Easton, David, 98
Eisenberg, David, 119, 121
electrical installation, 111–114
embodied carbon dioxide equivalent, 5–6, 7
embodied energy, 5–6, 7
end panels, 90–92, *91*
environmental product declarations (EPDs), 6, 31
equipment
    cost estimates, 101, *101,* 102, *103*
    loading, 76–77
    miscellaneous, 80
    mixing, 72–74
    sieving, 74
    tamping, 78
erosion control, 12
expansive clays, 24, 27
exterior sealers, 116–117

F

feature walls, 57–58
fiber reinforcement, 30
fiberglass grids, 59
field tests, 35–39
finishes
    exterior, 116–117
    interior, 110–111, 114–116
fly ash, 30
foam glass, 60
footings, 54, 60
force, 62
formply, 81–82
    See also formwork
formwork
    components, *88, 91*
    connectors, 96
    deflection limit, 87
    difference from concrete forms, 81
    end panels, 90–92
    formply, 81–82
    horizontal pressure on, 86
    importance of good design, 99–100
    insulation, 90
    outside corners, 92, 94–95, *94*
    patterned, 115
    removal, 99, 128
    sources of, 80
    strong-backs, 83, 88–89
    support of, 83–85
    VDB placement, 95–96
    walers, 85–86, 87–88
foundations, 54–56
France, organizations, 123

G

garden walls, 57–58
Germany, standards and codes, 122
glass fiber reinforced polymer (GFRP) rod, 59
grain-size distribution test, 45
GUL (ground-up limestone) Portland cement, 30

H

Hammer, Martin, 121
HB 195, 122
heating system, 80
Heyman, Jacques, 62
high-density polyethylene (HDPE), 59
horizontal shaft mixers, 74
H/T ratio, 63
humidity, effect of, 42
humidity buffering, 16–17
hydrated lime, 3, 31
hydronic pipes, 18

I

illite, 27
Imperial measurement system, 62
India, organizations, 123
innovations, 19
insect issues, 12
insulation
    for basements, 9
    double-wythe, 58–60
    external, 53, 54
    in formwork, 90
    internal, 55, 56
    requirements, 19
    as thermal control layer, 18
interior finishes, 110–111, 114–116
interior sealers, 114–115
International Residential Code, 120–121
Inventory of Carbon & Energy (ICE), 5
isotropic, 43

J

jar test, 39–41, *40*

K

kaolinite, 27
Keable, Julian, 28
Keable, Rowland, 28
Keable & Keable Lean Mix, *28*
Keable & Keable Rich Mix, *28*

King, Bruce, 30, 31, 121
Krayenhoff, Meror, 53

L

labor
    cost estimates, 101, *101*, 102
    experience of crew, 100
    roles of crew, 99
laboratory tests, 45–48
large volume projects, mixing procedure, 73–74
laser levels, 80
Lehmbau Regeln (Germany), 122
lifts, procedure, 97–98
Limestone Calcined Clay Cement (LC3), 30
lintel design, 69
load capacity, 64–65
loading equipment, 76–77

M

Making Better Concrete (King), 30, 31
manual tampers, 78
Martirena, Fernando, 30
masonry design codes, 4–5
mass, 62
materials
    cost estimates, 101–103, *101, 102, 103*
    embodied energy and carbon, 7
    See also specific materials
matric suction, 3
measurements
    definitions, 62
    weight vs volume, 33
mechanical reinforcement, 60
Metapor, 31
Middleton, G.F., 122
mixes
    ingredient measurements, 71
    large volume procedure, 73–74
    lift size, 97
    particle size proportions, 27–30
    small volume procedure, 72–73, 74
mixing equipment, 72–74

mixing pads, 72
monolithic walls, 53–58
montmorillonite, 24, 27
mortar mixers, 74
MSJC: Building Code Requirements for Masonry Structures, 63, 67

N

nails, 96
National Building Code of Canada (NBCC), 120
The New Carbon Architecture (King), 30
New Mexico Earthen Building Materials Code, 123
New Zealand, standards and codes, 122
Nez, George, 1–2
nibble/taste test, 35
nonsymmetrical double-wythe walls, 60, *61*
nonvertical loading, 66–67, *66*
NZS 4297:1998 Engineering Design of Earth Buildings, 63, 68–69, 122
NZS 4298:1998 Materials and Workmanship for Earth Buildings, 122
NZS 4299:1998 Earth Buildings Not Requiring Specific Design, 122

O

Ontario Building Code, 122
openings
    doors, 105
    windows, 105–106, *106, 107*, 108–109
optimum moisture content test, 45
out-of-plane loading, 65, *65*
outside corners, 92, 94–95, *94*
oxides, 31, 33, 71, 73

P

paint mixers, 73
penetrations, 114
perms, 16
personal protective equipment (PPE), 72, 98
pigments, 31, 33, 71, 73

piles, 54–55
Pima County, Arizona, 123
pipe clamps, 91
plasticity, 26–27
Plasticure, 32
plumbing, 114
pneumatic tampers, 78, 97
Poraver, 31
Portland cement, 3, 30
pozzolans, 3
   See also binders
pressure, 62
prisms, 44, 46–48, *47, 48*
Proctor test, 42

R
rainscreen, 12
rammed earth, definition, 3
rammed earth buildings, advantages of, 1, 19
   See also raw rammed earth; stabilized rammed earth
Rammed Earth Structures (Keable and Keable), 28
Rammed Earth (Walker et al.), 27–28
raw rammed earth
   construction phase tests for, 48–51
   design guidelines, 62–69
   disadvantages of, 3–4
   effect of aggregate mix, 34
   embedded bolts, 68, *68*
   field tests for, 35–39
   monolithic walls, 53–56
   particle size proportions, 27–29
   shop tests for, 39–45
   thickness, 53, 63
   work flow, 99
rebar, 59, 60
regulations. *See* building codes
reinforcement, 60
renovations, 117
repairs, 117
ribbon test, 39

roll test, 37–38, *38*
roof system, connection to, 60
rototillers, 72

S
safety equipment, 72, 98
safety regulations, enclosed space depth, 98
sample alternative solutions proposal, 131–132
sample engineering specification, 125–129
screws, 96
sealers, 4, 31–32, 73, 115–117
sensitive clays, 24
shear stress, definition, 65
shop tests, 39–45
shrink box test, 42–44, *43, 44*
SI system, 62
sieving, 74
silica emulsion sealers, 4, 32, 73, 116–117
sill plates, 109, 110
skid steers
   loading, 76–77
   mixing procedure, 73–74
slap test, 36–37, *37*
small volume projects
   loading equipment, 76
   mixing procedure, 72–73, 74
smell test, 35
snap-ties, 83–85
soils
   characteristics of, 43
   classifications, 21–22, *22*
   grain size distribution, 23
   particle size proportions, 27–30
specifications sample document, 125–129
stabilized rammed earth
   ball test, 74
   construction phase tests for, 48–51
   cost estimates, 101–103
   definition, 3
   design guidelines, 69
   double-wythe walls, 58–60
   effect of aggregate mix, 35

field tests for, 36
particle size proportions, 29–30
shop tests for, 39–41, 44–45
work flow, 99
stabilizers. *See* binders
stack effect, *14, 15*
Standards Australia, 122
Standards New Zealand, 122
The Stone Skeleton (Heyman), 62
stress, definition, 65
stresses, 65–67
strong-backs, 83, 88–89, *88*
structural design guidelines
raw rammed earth, 62–69
stabilized rammed earth, 69
symmetrical double-wythe walls, 58–59, *59*

T

tampers, 97
tamping equipment, 78
tar, 3
Tarion standard, 87
tarps, 80
taste test, 35
Tech-Dry, 32
telehandlers, 76–77
temperature, effect of, 42
tensile stress, definition, 65
tensile stresses, 65–67
test walls, 44–45
tests
for aggregate mixes, 34–35
construction phase, 48–51
field, 35–39
laboratory, 45–48
shop, 39–45
thermal control layer, 18
thermal isolation, 60
thermal mass, 18
through-ties, 83, 85
TMS 402/ACI 530, 69
trowel test, 37

U

Unified Soil Classification System (USCS), 21, 22
United States, standards and codes, 123
utilities, 111–114

V

vacuum cleaners, 80
vapor barriers, 16–17, 55
vapor control layer, 16–17
vegetation issues, 12
ventilation, 13
Volume Displacement Blocks (VDBs), 95–96
volume measurements, 33
volunteers. *See* crews/volunteers

W

walers, 83, 85–86, 87–88
Walker, et al. Lean Mix, *29*
Walker, et al. Rich Mix, *29*
Walker, Pete, 27–28, 122
wall assemblies
double-wythe, 58–60, *59, 61*
finishing top, 57–58, 109–110
length, 67
maximum height, 63
monolithic, 53–56
types of, 53
wall samples, 44–45
wash test, 38–39
water control layer, 11–12
weather barrier, 11–12
weight, 62
weight measurements, 33
wind stress, 65, *65*
window openings, 105–106, *106, 107*, 108–109
winter conditions, protection of work, 80
work area
ingredient measurements, 71
mixing area setup, 72
work flow, 99
wythes, 12
See also wall assemblies

About the Author

TIM KRAHN is a registered professional engineer and partner in Building Alternatives Inc. (BAI). He holds a Master's degree in geotechnical engineering and a Bachelor's degree in civil engineering from the University of Manitoba. He has experience in residential and commercial construction and holds a certificate in carpentry and woodworking from Red River College. Tim is also a LEED™ accredited professional, an active member of the Timber Frame Guild's Timber Frame Engineering Council, and a founding member of the Natural Buildings Engineering Group. Tim currently sits on the board of directors for the Ontario Natural Building Coalition.

BAI is a consulting engineering firm licensed in BC, Alberta, Saskatchewan, Manitoba, Ontario, and the Northwest Territories and Nunavut. BAI actively participates in educational programs both at university and college level institutions as well as through professional and trade associations. BAI is also active in applied research, working together with academic institutions on projects to improve our quantitative knowledge of pre-industrial building materials and their behavior in the Canadian climate, working with the Alternative Village at the University of Manitoba and with researchers at Queen's University, University of New Brunswick, and others.

BAI specializes in engineering design using low-embodied energy, low carbon emission materials such as wood, earth, stone, straw, clay, hemp, and lime. This engineering design includes the techniques with which these materials are combined and assembled, and incorporates responsible and appropriate current technologies such as air barrier membranes and

geosynthetics that allow new applications of the older, time tested materials.

Tim and his partner Dalila have lived in southeastern Ontario since 2010, moving from Manitoba to take care of their family's property. While in Winnipeg, Tim was active in the inner city housing community as well as at the University of Manitoba, where he was a sessional and technical instructor in both architecture and engineering. Tim also spent two years as the coordinator of the Alternative Village at the U of M, which was founded by Building Alternatives principal, Dr. Kris Dick. Tim's research interests include earthen construction, energy and material efficiency, and sustainability in the built environment.

A Note About the Publisher

NEW SOCIETY PUBLISHERS is an activist, solutions-oriented publisher focused on publishing books for a world of change. Our books offer tips, tools, and insights from leading experts in sustainable building, homesteading, climate change, environment, conscientious commerce, renewable energy, and more — positive solutions for troubled times.

We're proud to hold to the highest environmental and social standards of any publisher in North America. This is why some of our books might cost a little more. We think it's worth it!

- We print all our books in North America, never overseas
- All our books are printed on **100% post-consumer recycled paper**, processed chlorine free, with low-VOC vegetable-based inks (since 2002)
- Our corporate structure is an innovative employee shareholder agreement, so we're one-third employee-owned (since 2015)
- We're carbon-neutral (since 2006)
- We're certified as a B Corporation (since 2016)

At New Society Publishers, we care deeply about *what* we publish — but also about how we do business.

Download our catalogue at https://newsociety.com/Our-Catalog or for a printed copy please email info@newsocietypub.com or call 1-800-567-6772 ext 111

New Society Publishers
ENVIRONMENTAL BENEFITS STATEMENT

For every 5,000 books printed, New Society saves the following resources:[1]

36	Trees
3,243	Pounds of Solid Waste
3,568	Gallons of Water
4,654	Kilowatt Hours of Electricity
5,895	Pounds of Greenhouse Gases
25	Pounds of HAPs, VOCs, and AOX Combined
9	Cubic Yards of Landfill Space

[1] Environmental benefits are calculated based on research done by the Environmental Defense Fund and other members of the Paper Task Force who study the environmental impacts of the paper industry.